Susanne Westhoff

Probes of Yukawa Unification in Supersymmetric SO(10) Models

Susanne Westhoff

Probes of Yukawa Unification in Supersymmetric SO(10) Models

Constraining the flavour structure of Yukawa corrections

Südwestdeutscher Verlag für Hochschulschriften

Impressum/Imprint (nur für Deutschland/ only for Germany)
Bibliografische Information der Deutschen Nationalbibliothek: Die Deutsche Nationalbibliothek verzeichnet diese Publikation in der Deutschen Nationalbibliografie; detaillierte bibliografische Daten sind im Internet über http://dnb.d-nb.de abrufbar.
  Alle in diesem Buch genannten Marken und Produktnamen unterliegen warenzeichen-, marken- oder patentrechtlichem Schutz bzw. sind Warenzeichen oder eingetragene Warenzeichen der jeweiligen Inhaber. Die Wiedergabe von Marken, Produktnamen, Gebrauchsnamen, Handelsnamen, Warenbezeichnungen u.s.w. in diesem Werk berechtigt auch ohne besondere Kennzeichnung nicht zu der Annahme, dass solche Namen im Sinne der Warenzeichen- und Markenschutzgesetzgebung als frei zu betrachten wären und daher von jedermann benutzt werden dürften.

Verlag: Südwestdeutscher Verlag für Hochschulschriften Aktiengesellschaft & Co. KG
Dudweiler Landstr. 99, 66123 Saarbrücken, Deutschland
Telefon +49 681 37 20 271-1, Telefax +49 681 37 20 271-0
Email: info@svh-verlag.de
Zugl.: Karlsruhe, KIT, Diss., 2009

Herstellung in Deutschland:
Schaltungsdienst Lange o.H.G., Berlin
Books on Demand GmbH, Norderstedt
Reha GmbH, Saarbrücken
Amazon Distribution GmbH, Leipzig
ISBN: 978-3-8381-1307-4

Imprint (only for USA, GB)
Bibliographic information published by the Deutsche Nationalbibliothek: The Deutsche Nationalbibliothek lists this publication in the Deutsche Nationalbibliografie; detailed bibliographic data are available in the Internet at http://dnb.d-nb.de.
  Any brand names and product names mentioned in this book are subject to trademark, brand or patent protection and are trademarks or registered trademarks of their respective holders. The use of brand names, product names, common names, trade names, product descriptions etc. even without a particular marking in this works is in no way to be construed to mean that such names may be regarded as unrestricted in respect of trademark and brand protection legislation and could thus be used by anyone.

Publisher: Südwestdeutscher Verlag für Hochschulschriften Aktiengesellschaft & Co. KG
Dudweiler Landstr. 99, 66123 Saarbrücken, Germany
Phone +49 681 37 20 271-1, Fax +49 681 37 20 271-0
Email: info@svh-verlag.de

Printed in the U.S.A.
Printed in the U.K. by (see last page)
ISBN: 978-3-8381-1307-4

Copyright © 2010 by the author and Südwestdeutscher Verlag für Hochschulschriften Aktiengesellschaft & Co. KG and licensors
All rights reserved. Saarbrücken 2010

Das Wort "Theorie" stand einmal für das Vermögen, in einem alles und alles in einem zu sehen. Wer jedoch das Eine nicht zu schauen vermag (und manches spricht dafür, dass niemand es kann), darf sich an vieles halten. So geschieht es hier.

Martin Seel

Contents

Introduction   3

1 Elements of Unification   7
   1.1 Flavour of and in the Standard Model 7
   1.2 Supersymmetric flavour physics 13
   1.3 Supersymmetric Yukawa unification 17
   1.4 Embedding the Standard Model into SO(10) 22
   1.5 SO(10) breaking 24
   1.6 Yukawa unification in SO(10) 25
   1.7 Higher-dimensional Yukawa terms 27

2 A Supersymmetric SO(10) Model of Flavour   31
   2.1 The Chang-Masiero-Murayama model 31
   2.2 Constraints on the parameter space 35
   2.3 Phenomenology of down-quark-lepton unification 39
   2.4 Numerical setup 47
   2.5 The CP phase in $B_s - \bar{B}_s$ meson mixing 49

3 Yukawa Corrections for Light Fermions   51
   3.1 Corrections from higher-dimensional operators 51
   3.2 Constraints from ϵ_K 55
   3.3 Constraints from $B_d - \bar{B}_d$ and $B_s - \bar{B}_s$ mixing 57
   3.4 Closing the unitarity triangle 58
   3.5 The flavour structure of Yukawa corrections 62

4 Supersymmetric Unification and Large $\tan\beta$ 65
   4.1 Effects of large $\tan\beta$ in flavour physics 65
   4.2 Charged Higgs in $B \to (D)\tau\nu$ branching ratios 68
   4.3 $B \to D\tau\nu$ differential distributions 74
   4.4 Yukawa unification and $\tan\beta$ 79

Conclusions   81

A	**Appendix**	**85**
	A.1 Weyl spinors	85
	A.2 SO(10) decompositions	86
	A.3 Loop functions for meson mixing	87
	A.4 $B \to D$ form factors and decay distribution	88

Bibliography   89

Acknowledgements   101

Introduction

Among the grand open questions any (curious) human being has in mind are the following

Why do we observe so many things we do not understand?

Is there a hidden order beyond our reach of perception?

(How) does this affect our daily living?

For a contemporary particle physicist, these questions turn into quests

Why do we observe different elementary particles and forces?

Is there a fundamental theory at very high energies?

How can we probe ultra-high-energy physics at today's particle colliders?

The persistence of the questions reflects the difficulty, if not impossibility, to find solid answers. Still (or hence), the remarkable achievements in particle physics within the last century and the promising discovery potential of the Large Hadron Collider (LHC) encourage scientists to proceed in approaching "the truth". It is a worthwhile attempt to understand the depth of the open issues and to subsequently contribute to find solutions.

Particles and forces

The guiding principle to explain the properties and interactions of elementary particles are symmetries. A physical law based on symmetry has, on top of descriptive, predictive power. What is established today as the Standard Model of particle physics is a quantum field theory with an $SU(3) \times SU(2) \times U(1)$ gauge symmetry of strong, weak, and electromagnetic interactions. This framework describes the interactions between fermions (quarks and leptons) via the exchange of gauge bosons (gluons, W and Z bosons, and the photon). Exploring the symmetry properties of these interactions allows to classify the quarks and leptons in three generations. The suppression of flavour-changing neutral currents in weak interactions led to the prediction of the charm quark to complete the second generation of quarks. In order to explain the observed CP violation in weak interactions of neutral kaons, the existence of a third generation was postulated and confirmed by the subsequent discovery of the tau lepton and the bottom and top quarks. Also in the lepton sector, the gauge symmetry of the Standard Model properly describes the electroweak interactions of charged electrons, muons, and tau leptons with the corresponding neutrinos. This extremely successful symmetry picture, however, forbids elementary mass terms in the Lagrangean. In the Standard Model, the generation of mass is implemented through the Higgs mechanism. The vacuum expectation value of an additional scalar field, the Higgs field, spontaneously breaks the SU(2) × U(1) symmetry of electroweak interactions. This procedure leads to massive W and Z bosons in the range of the electroweak scale $M_{\text{EW}} \simeq 100\,\text{GeV}$. Interactions of the Higgs field with the fermions, so-called Yukawa couplings, generate masses for quarks and leptons.

A fundamental theory at high energies

The Standard Model is experimentally confirmed to describe particle interactions up to energies of the order of M_{EW} with impressive accuracy. Its conceptual validity, however, covers the vast range up to a high-energy end marked by the Planck scale $M_{\text{Pl}} = 10^{19}\,\text{GeV}$, where gravitational interactions are relevant. Strong, weak, and electromagnetic interactions are experimentally proven to be energy-dependent. The gauge couplings associated with strong and weak interactions decrease at high energies, while the electromagnetic coupling increases. The energy dependence is well described by the theoretical concept of renormalization group evolution and applies to all couplings in the theory, including the Yukawa couplings.

The assumption of a larger symmetry of particle interactions at high energies is not only temptingly beautiful but also mathematically substantiated. Within a Grand Unified Theory (GUT) based on the symmetry groups SU(5) or SO(10) the strong, weak, and electromagnetic couplings are predicted to merge at an intermediate scale $M_{\text{GUT}} \simeq 10^{15}\,\text{GeV}$. Three of the fundamental forces thereby merge into one. The Standard Model is contained in the larger symmetry group and can be considered as the low-energy remnant of a more fundamental theory. Accordingly, the Standard-Model quarks are embedded together with the leptons into representations of SU(5) or SO(10). On these grounds one can explain important issues left open by the Standard Model like the quantization of electric charges or the number of fermions per generation. The unification of fermions induces new interactions between quarks and leptons via the exchange of a heavy boson associated with the larger gauge symmetry. These interactions induce proton decay, which poses a serious challenge to GUTs, since it has not been observed in experiments so far.

Independent from the embedding into a Grand Unified Theory, the large energy range of the Standard Model is problematic. Interactions of the Higgs field with heavy particles generally destabilize the scale of electroweak symmetry breaking, which is known as the hierarchy problem. A promising way to protect the Higgs mass from unnaturally large corrections is the introduction of supersymmetry, a symmetry between bosons and fermions. By assigning each fermionic degree of freedom of the Standard Model a bosonic superpartner (and vice versa), Higgs mass corrections due to couplings to heavy particles are cancelled by contributions of the corresponding superpartners. In order to provide an effective solution to the hierarchy problem, one expects supersymmetric particles not far

Introduction   5

above the electroweak scale, i.e. at around $M_{\text{SUSY}} \simeq 1\,\text{TeV}$.
As a fortunate coincidence Grand Unification likes supersymmetry. The presence of superpartners can prolongate the lifetime of the proton, which temporarily reconciles the idea of GUTs with the non-observation of its decay. Moreover, Grand Unification needs supersymmetry. Numerically, the measured gauge couplings unify only if effects of superparticles in the evolution to high energy scales are taken into account. Besides making the unification of gauge couplings viable, supersymmetry allows to probe a second aspect of Grand Unification, the unification of Yukawa couplings. If quarks and leptons of all three generations interact with the Higgs field via one and the same Yukawa coupling, flavour mixing between quarks translates into flavour-changing neutral currents among leptons and vice versa. Such relations cannot be observed in Standard-Model currents, but become visible in interactions involving superpartners. From experiment we know that flavour mixing in the quark sector is moderate. Contrarily, the large angles measured in solar and atmospheric neutrino oscillations show that mixing in the lepton sector is significant. In a Grand Unified Theory, left-handed charged leptons and neutrinos are embedded together with right-handed down-type quarks and their superpartners into one representation of SU(5). The large neutrino mixing can thereby as well appear in neutral currents with right-handed down-type (s)quarks. Signatures of quark-lepton unification at high energies can thus be read off in flavour physics observables at low energies. In particular, the large atmospheric neutrino mixing angle enters bottom-strange transitions, yielding sizeable effects in observables related to B_s mesons. The current precision of B_s observables leaves room for signatures of new physics beyond the Standard Model, whose observation in the near future can support the idea of Grand Unification.

Ultra-high-energy physics at today's particle colliders

Symmetries are very useful guides to the main features of particle interactions, but nature is more intricate. Perfect Yukawa unification complies with experiment only for the third generation, i.e. for top and bottom quarks and tau leptons. A realistic GUT model of fermion masses and mixings requires a refinement of the Yukawa sector. In this work we confront the imprints of unified quarks and leptons with measured flavour physics observables to gain insight into the structure of Yukawa couplings. We focus on the corrections to Yukawa unification for light fermions. They translate atmospheric neutrino mixing also into strange-down and bottom-down transitions. The flavour structure of Yukawa corrections is strongly constrained from accurate predictions and measurements of kaon observables, as well as from B_d physics. Extensions of the Yukawa sector are thereby restricted to exhibit a specific form, which is an important finding for GUT model building.
Constraining the parameters of a concrete GUT model from sensitive flavour observables helps to discriminate between different avenues of unification. We make further use of this method to gain information about the Higgs sector in supersymmetry that comprises five Higgs bosons, two of which carry electric charge. Its key parameter $\tan\beta$ determines the ratio of top and bottom Yukawa couplings. A consistent GUT model with top-bottom-tau unification requires $\tan\beta$ to be large. Determining the magnitude of $\tan\beta$ is therefore

crucial to figure out the overall Yukawa structure realized in a framework with Grand Unification. Large $\tan\beta$ induces characteristic effects in flavour physics observables, which we exploit to pin down its value. In particular, the couplings of charged Higgs bosons to bottom quarks and tau leptons are enhanced if $\tan\beta$ is large. The (semi)leptonic meson decay modes $B \to \tau\nu$ and $B \to D\tau\nu$ thus strongly depend on $\tan\beta$. We point out that the differential decay distributions in $B \to D\tau\nu$ are well suited to discriminate between Standard-Model and charged-Higgs contributions through the shape of the spectrum. The benefit of discovering charged-Higgs effects in $B \to (D)\tau\nu$ decays at the B factories is twofold: First, it would confirm an extended Higgs sector conform with supersymmetry. Second, for a fixed spectrum of supersymmetric particles, the measurement of $\tan\beta$ could clarify fundamental aspects of Yukawa unification.

This work is composed as follows: In Chapter 1, the disposed reader is made familiar with the foundations of flavour physics and Grand Unification, including group-theoretical aspects of SO(10). In Chapter 2, we introduce a specific supersymmetric GUT model based on SO(10) and designed to probe down-quark-lepton Yukawa unification. Within this framework we explore the effects of large atmospheric neutrino mixing in bottom-strange transitions on the mass difference and CP phase in $B_s - \overline{B}_s$ meson mixing. Chapter 3 is devoted to corrections to Yukawa unification. We derive constraints on Yukawa corrections for light fermions from $K - \overline{K}$ and $B_d - \overline{B}_d$ mixing. As an application we study implications of neutrino mixing effects in CP-violating K and B_d observables on the unitarity triangle. Finally, in Chapter 4, we discuss effects of large $\tan\beta$ in $B \to (D)\tau\nu$ decays with respect to their potential to discover charged Higgs bosons and to discriminate between different GUT models of flavour. The covered subjects are to a large extent published in Refs. [1] and [2].

Chapter 1

Elements of Unification

Although Grand Unification (if existing) is realized at energy scales far beyond today's colliders' reach, it can leave characteristic imprints in low-energy flavour physics observables. The analysis of flavour effects in GUTs requires a framework that is valid over a large energy range up to the Planck scale. We will start by introducing crucial aspects of flavour physics in the Standard Model, before extending the setup by supersymmetry. The low-energy theory being settled, we move on to implications of high-energy Yukawa unification on flavour mixing. The subsequently presented mathematical grounds for SO(10) unification and breaking are completed with an introduction to higher-dimensional Yukawa terms.

1.1 Flavour of and in the Standard Model

In order to embed the Standard Model into a theory valid up to high scales, it is important to control the low-energy basis we start from. In fact, many formal features of Grand Unification are visible in the mathematical structure of Standard-Model particle interactions. The Standard Model itself can be seen as an example of unification: It combines weak and electromagnetic interactions in the framework of one unified gauge theory, described by an SU(2) × U(1) symmetry [3–5]. Expanding this group structure by quantum chromodynamics, one ends up with the gauge group of the Standard Model

$$G_{\text{SM}} = \text{SU}(3)_C \times \text{SU}(2)_L \times \text{U}(1)_Y \,, \tag{1.1}$$

incorporating strong, weak, and electromagnetic interactions between elementary particles. The gauge group G_{SM} acts on three generations of fifteen distinct fermion fields each, written down in the upper part of Tab. 1.1. Left-(right-)handed fermions are denoted by an index $L(R)$, and $\nu_{e,\mu,\tau}$ represent neutrinos. Under the gauge group G_{SM} these fermions transform as $(R_3, R_2)_Y$, with R_3 and R_2 labelling the representations of SU(3) and SU(2). Only the quarks take part in strong interactions, since they transform as triplets $R_3 = 3$ under SU(3)$_C$. Further, only left-handed particles are weakly interacting, as they come in SU(2)$_L$ doublets $R_2 = 2$, namely $Q = (u, d)_L$ and $L = (\nu, e)_L$. Apart from the neutrinos,

quarks and leptons	$(u = \text{up},\quad d = \text{down})_L$ $(c = \text{charm},\ s = \text{strange})_L$ $(t = \text{top},\quad b = \text{bottom})_L$	u_R c_R t_R	d_R s_R b_R	$(\nu_e, e)_L$ $(\nu_\mu, \mu)_L$ $(\nu_\tau, \tau)_L$	e_R μ_R τ_R
$(R_3, R_2)_Y$	$(3,2)_{1/3}$	$(3,1)_{4/3}$	$(3,1)_{-2/3}$	$(1,2)_{-1}$	$(1,1)_{-2}$
Q_e [e]	$(2/3, -1/3)$	$2/3$	$-1/3$	$(0, -1)$	-1

Table 1.1: Fermions of the Standard Model, their transformation properties under the gauge group G_{SM}, and electric charges Q_e in units of the electron charge e.

bosons	g^a	$W^{1,2}, W^3$	B	$H = (H^+, H^0)$
$(R_3, R_2)_Y$	$(8,1)_0$	$(1,3)_0$	$(1,1)_0$	$(1,2)_1$
Q_e [e]	0	$(\pm 1, 0)$	0	$(1, 0)$

Table 1.2: Bosons of the Standard Model.

all fermions experience electromagnetic interactions, manifest in their electric charge Q_e. The quantum number corresponding to U(1), the hypercharge Y, is linked to Q_e via the three-component of weak isospin, $T_3 = \pm 1/2$, as $Q_e = T_3 + \frac{1}{2}Y$. Particle interactions are described by the exchange of force carriers, the gauge bosons, depicted in Tab. 1.2. Strong interactions are mediated by a set of eight gluons g^a with $a = 1, \ldots, 8$; electroweak interactions involve three gauge bosons of SU(2)$_L$, W^a with $a = 1, 2, 3$, and one gauge boson B associated with U(1)$_Y$.

The gauge group G_{SM} is well suited to describe interactions; however, it forbids the introduction of particle masses. What became the "standard" way out of this striking inconvenience is known as the Higgs mechanism [6–9]: The symmetry of electroweak interactions is broken spontaneously down to a remnant associated with electromagnetism. Spontaneous symmetry breaking solely affects the vacuum state of the gauge theory, whereas the dynamical properties continue to follow G_{SM}. To break electroweak symmetry spontaneously, one adds a complex scalar SU(2) doublet, the Higgs field H (see Tab. 1.2), which exhibits a vacuum expectation value (vev) $v = 174\,\text{GeV}$ in its electrically neutral component, so that

$$\text{SU}(2)_L \times \text{U}(1)_Y \xrightarrow{v} \text{U}(1)_{\text{em}}\,; \qquad H = \begin{pmatrix} H^+ \\ H^0 \end{pmatrix}, \ \langle H \rangle = \begin{pmatrix} 0 \\ v \end{pmatrix}. \qquad (1.2)$$

The three degrees of freedom associated with electroweak symmetry breaking translate into longitudinal modes of the gauge bosons $W^\pm$ and Z, thereby giving them a mass. The

1.1. Flavour of and in the Standard Model

fourth degree of freedom associated with U(1)$_{\text{em}}$, the photon A, remains massless:

$$W^{\pm} = (W^1 \mp i W^2)/\sqrt{2}, \qquad M_W^2 = g_2^2 v^2/2$$
$$Z = \cos\theta_w W^3 - \sin\theta_w B, \qquad M_Z^2 = (g_1^2 + g_2^2) v^2/2 \qquad (1.3)$$
$$A = \sin\theta_w W^3 + \cos\theta_w B, \qquad M_A = 0.$$

The relative strength between the SU(2) and U(1) couplings g_1 and g_2 is parametrized by the Weinberg angle $\theta_w = \arctan(g_1/g_2)$. The masses of gauge bosons and fermions result from couplings to the Higgs field.

Flavour mixing

Higgs-fermion couplings are summarized in the Yukawa sector of the Standard Model,

$$\mathcal{L}_{\text{SM}}^Y = -Y_u^{ij} Q_i u_j^c H + Y_d^{ij} Q_i d_j^c H^* + Y_e^{ij} L_i e_j^c H^* + \text{h.c.}, \qquad (1.4)$$

where the fermion fields $Q, u^c, \ldots$ are left-handed Weyl spinors. Details of the Dirac structure are suppressed, but given in Appendix A.1. Superscripts c indicate charge conjugation, under which right-handed fermions transform into the corresponding left-handed anti-fermions (and vice versa). The SU(2) doublets Q and L are understood to couple to the Higgs doublet in a gauge-invariant way, e.g.[1]

$$Q u^c H = Q_\alpha u^c \epsilon^{\alpha\beta} H_\beta \qquad \text{with} \quad \epsilon^{12} = -\epsilon^{21} = -\epsilon_{12} = 1. \qquad (1.5)$$

The Yukawa couplings Y_f^{ij} are 3×3 matrices in flavour space, the indices $i, j = 1, 2, 3$ denoting the respective fermion generation. Fermion masses M_f are obtained by setting $H \to \langle H \rangle$ and after diagonalizing the Yukawa matrices Y_f,

$$M_u = v \cdot \begin{pmatrix} y_u & 0 & 0 \\ 0 & y_c & 0 \\ 0 & 0 & y_t \end{pmatrix}, \quad M_d = v \cdot \begin{pmatrix} y_d & 0 & 0 \\ 0 & y_s & 0 \\ 0 & 0 & y_b \end{pmatrix}, \quad M_e = v \cdot \begin{pmatrix} y_e & 0 & 0 \\ 0 & y_\mu & 0 \\ 0 & 0 & y_\tau \end{pmatrix}. \qquad (1.6)$$

Yukawa diagonalization is achieved by two unitary matrices $L_f, R_f : Y_f = L_f^* \hat{Y}_f R_f^\top$. These matrices simultaneously rotate the fermion fields into their mass eigenstates,

$$u_L = L_u (u_L)^m, \qquad u_R = R_u (u_R)^m, \qquad \nu_L = L_\nu (\nu_L)^m,$$
$$d_L = L_d (d_L)^m, \qquad d_R = R_d (d_R)^m, \qquad e_L = L_e (e_L)^m, \qquad e_R = R_e (e_R)^m. \qquad (1.7)$$

Within the Standard Model, only two combinations of the rotations of left-handed fermions are physical,[2] namely

$$V_{\text{CKM}} \equiv L_u^\dagger L_d = \begin{pmatrix} V_{ud} & V_{us} & V_{ub} \\ V_{cd} & V_{cs} & V_{cb} \\ V_{td} & V_{ts} & V_{tb} \end{pmatrix} \quad \text{and} \quad V_{\text{PMNS}} \equiv L_e^\dagger L_\nu = \begin{pmatrix} V_{e1} & V_{e2} & V_{e3} \\ V_{\mu 1} & V_{\mu 2} & V_{\mu 3} \\ V_{\tau 1} & V_{\tau 2} & V_{\tau 3} \end{pmatrix}, \qquad (1.8)$$

---
[1] $\alpha, \beta = 1, 2$ are SU(2) indices.
[2] Here we assume massive light neutrinos to introduce notations for later convenience. Strictly speaking, in the Standard Model, $m_\nu = 0$, so that $L_{e,\nu}$ and consequently V_{PMNS} are not physical.

named after their "fathers" Cabibbo, Kobayashi, Maskawa [10, 11] and Pontecorvo, Maki, Nakagawa, Sakata [12, 13]. These are unitary 3 × 3 matrices parametrized by three angles and six phases each. Five phases can be eliminated by a redefinition of the fermion fields. A convenient expression of the remaining, physical degrees of freedom in V_{CKM} and V_{PMNS} is provided by the standard parametrization in terms of three angles θ_{12}, θ_{23}, θ_{13} and one phase δ,[3]

$$V_{\text{CKM,PMNS}} = \begin{pmatrix} c_{12}c_{13} & s_{12}c_{13} & s_{13}e^{-i\delta} \\ -s_{12}c_{23} - c_{12}s_{23}s_{13}e^{i\delta} & c_{12}c_{23} - s_{12}s_{23}s_{13}e^{i\delta} & s_{23}c_{13} \\ s_{12}s_{23} - c_{12}c_{23}s_{13}e^{i\delta} & -c_{12}s_{23} - s_{12}c_{23}s_{13}e^{i\delta} & c_{23}c_{13} \end{pmatrix}. \quad (1.9)$$

These matrices describe flavour mixing in the quark and lepton sectors, visible in weak couplings to charged gauge bosons,[4]

$$\begin{aligned}\mathcal{L}^{\text{CC}} &= -\frac{g_2}{\sqrt{2}} \left(u_L^c \bar{\sigma}^\mu d_L W_\mu^+ + \nu_L^c \bar{\sigma}^\mu e_L W_\mu^+ \right) + \text{h.c.} \\ &= -\frac{g_2}{\sqrt{2}} \left((u_L^c)^{\text{m}} V_{\text{CKM}} \bar{\sigma}^\mu (d_L)^{\text{m}} W_\mu^+ + (\nu_L^c)^{\text{m}} V_{\text{PMNS}}^\dagger \bar{\sigma}^\mu (e_L)^{\text{m}} W_\mu^+ \right) + \text{h.c.} \, . \end{aligned} \quad (1.10)$$

Note that charged gauge bosons couple only to left-handed fermions, which reflects the parity-breaking feature of weak interactions. Besides these charged currents, the gauge group of the Standard Model allows for electromagnetic and weak neutral currents,

$$\mathcal{L}^{\text{NC}} = -e\, Q_e(f)\, (f_L^c \bar{\sigma}^\mu f_L + f_R^c \sigma^\mu f_R) A_\mu - \frac{g_2}{\cos\theta_W} (g_L^f f_L^c \bar{\sigma}^\mu f_L + g_R^f f_R^c \sigma^\mu f_R)\, Z_\mu,$$
with $e = g_2 \sin\theta_W$, $g^f_{L,R} = T_3(f_{L,R}) - Q_e(f_{L,R}) \sin^2\theta_W$. $\quad (1.11)$

The Dirac structure of these currents is prescribed by $\sigma^\mu = (\sigma^0, \sigma^i)$ and $\bar{\sigma}^\mu = \sigma_\mu = \eta_{\mu\nu}\sigma^\nu$ with the metric $\eta_{\mu\nu} = \text{diag}(1, -1, -1, -1)$, $\sigma^0 = \mathbf{1}_2$, and the three generators of SU(2), the Pauli matrices σ^i,

$$\sigma^1 = \begin{pmatrix} 0 & 1 \\ 1 & 0 \end{pmatrix}, \quad \sigma^2 = \begin{pmatrix} 0 & -i \\ i & 0 \end{pmatrix}, \quad \sigma^3 = \begin{pmatrix} 1 & 0 \\ 0 & -1 \end{pmatrix}. \quad (1.12)$$

Neutral currents do not involve the flavour-mixing matrix V_{CKM}. In other words, there are no flavour-changing neutral currents (FCNC) in the Standard Model at tree level and, in the limit of vanishing quark masses, nor at loop level. This is known as the Glashow-Iliopoulos-Maiani (GIM) mechanism [5]. The strong suppression of FCNC in the Standard Model is advantageous for studying new-physics contributions in observables based on $b-s$, $b-d$, and $s-d$ transitions.

Contrarily to strong and neutral electroweak interactions, which are invariant under charge conjugation C, parity P, and time reversal T transformations,[5] charged weak interactions

---

[3]The abbreviations c_{ij} and s_{ij} stand for $\cos\theta_{ij}$ and $\sin\theta_{ij}$. Potential Majorana phases in V_{PMNS} are dropped, since they are not relevant in our analysis.
[4]Note that for $m_\nu = 0$, charged currents with leptons are flavour-conserving.
[5]Here we disregard the strong CP problem.

1.1. Flavour of and in the Standard Model

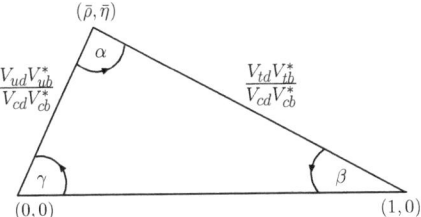

Figure 1.1: Unitarity triangle.

break C, P, and the combination CP. To make the last statement explicit, let us write down the transformation properties of vector currents and charged gauge bosons under CP,

$$CP\; u_L^c \bar{\sigma}^\mu d_L = -d_L^c \bar{\sigma}_\mu u_L, \qquad CP\; W_\mu^\pm = -W^{\mp,\mu}. \tag{1.13}$$

The Lagrangean from Eq. (1.10) describing charged currents then transforms under CP as

$$\begin{aligned}\mathcal{L}_q^{\text{CC}} &= -\frac{g_2}{\sqrt{2}} \left(u_L^c V_{\text{CKM}} \bar{\sigma}^\mu d_L W_\mu^+ + d_L^c V_{\text{CKM}}^\dagger \bar{\sigma}^\mu u_L W_\mu^- \right) \\ \xrightarrow{CP} &\; -\frac{g_2}{\sqrt{2}} \left(d_L^c V_{\text{CKM}}^\top \bar{\sigma}^\mu u_L W_\mu^- + u_L^c V_{\text{CKM}}^* \bar{\sigma}^\mu d_L W_\mu^+ \right).\end{aligned} \tag{1.14}$$

CP is thus conserved in charged currents only if $V_{\text{CKM}} = V_{\text{CKM}}^*$ or, equivalently, if there is no complex phase δ in the CKM matrix. In the Standard Model, V_{CKM} is the only source of flavour and CP violation. In supersymmetric models, additional sources of flavour and CP violation arise, with drastic phenomenological implications. A graphic way to quantify CP violation in the Standard Model is derived by exploiting the unitarity of the CKM matrix, which implies

$$V_{ud} V_{ub}^* + V_{cd} V_{cb}^* + V_{td} V_{tb}^* = 0. \tag{1.15}$$

Normalized to $V_{cd}V_{cb}^*$, the summands in this equation span a triangle in the complex plane, the "unitarity triangle" (UT) depicted in Fig. 1.1. By overconstraining its sides and angles α, β, and γ experimentally from B and K physics observables, one has realistic prospects to distinguish CP violation in the Standard Model from possible new-physics contributions. Due to the considerable effort made to measure the parameters in V_{CKM} and V_{PMNS}, the structure of Standard-Model flavour mixing is experimentally constrained, though it is difficult to motivate it on theoretical grounds. In the quark sector, especially the collaborations at the B factories BaBar and BELLE provide us with precise determinations of CKM elements. They reveal a strongly hierarchical mixing with small off-diagonal elements in particular when heavy quarks are involved, see Fig. 1.2 left. Concerning the lepton sector, due to the fact that neutrinos escape the detector, it is currently not possible to

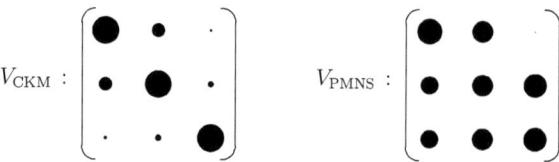

Figure 1.2: Quark and lepton mixing matrices. The absolute values of elements V_{ij} are represented by disks with radius $r \sim \sqrt{|V_{ij}|}$.

observe lepton flavour mixing in the decay products at colliders. Still, there are (looser) constraints on the PMNS angles θ_{12}, θ_{23}, and θ_{13} from solar, atmospheric, and reactor neutrino oscillation experiments. These measurements are in agreement with tri-bi-maximal mixing [14, 15], which corresponds to $\theta_{12} = \arcsin(1/\sqrt{3}) \simeq 35°$, $\theta_{23} = \arcsin(1/\sqrt{2}) = 45°$, and $\theta_{13} = 0°$, leading to

$$|V_{\text{PMNS}}| = \frac{1}{\sqrt{6}} \begin{pmatrix} 2 & \sqrt{2} & 0 \\ -1 & \sqrt{2} & \sqrt{3} \\ 1 & -\sqrt{2} & \sqrt{3} \end{pmatrix}. \quad (1.16)$$

In comparison with quark mixing, the structure of the lepton mixing matrix is much more "democratic", see Fig. 1.2 right. We would like to point out the large atmospheric neutrino mixing angle θ_{23}, which is close to maximal, while the "reactor" mixing angle θ_{13} is close to zero. Consequently, the third neutrino mass eigenstate is an (almost) equal mixture of ν_μ and ν_τ. This observation is particularly interesting in the framework of Grand Unification, where quarks and leptons are connected, so that the atmospheric mixing angle can induce large flavour mixing in the quark sector.

Neutrino masses

Lepton flavour mixing requires massive neutrinos and is thus a step beyond the Standard Model, where neutrinos are massless. Let us comment on the generation of light neutrino masses by interaction with heavy right-handed neutrinos, known as the seesaw mechanism [16–18]. Neutrino masses can be implemented by extending the Yukawa sector by a Standard-Model singlet $N = (1, 1)_0$,

$$\mathcal{L}^Y = \mathcal{L}^Y_{\text{SM}} + Y_\nu^{ij} L_i N_j H - \frac{1}{2} M_N^{ij} N_i N_j + \text{h.c..} \quad (1.17)$$

Heavy right-handed neutrinos are introduced as Majorana particles, i.e. they are their own antiparticles, $N = N^c$. Thus, by writing Eq. (1.17), we installed both Dirac and Majorana mass terms for neutrinos. If one considers the Standard Model as an effective theory, couplings of light to heavy neutrinos give rise to a dimension-five term in the low-energy Lagrangean,

$$\mathcal{L}_{\text{eff}} = -\frac{1}{2} L_i H \, (Y_\nu^{ik}(M_N^{-1})^{kl} Y_\nu^{lj}) \, L_j H + \text{h.c.} \,. \tag{1.18}$$

After electroweak symmetry breaking, this operator generates Majorana masses for light neutrinos,

$$\mathcal{L}^{M_\nu} = -\frac{1}{2} L_i \, M_\nu^{ij} \, L_j, \quad \text{with} \quad M_\nu^{ij} = v^2 \, Y_\nu^{ik}(M_N^{-1})^{kl} Y_\nu^{lj}. \tag{1.19}$$

To get light neutrino masses $M_\nu = \mathcal{O}(0.1)\,\text{eV}$, the mass scale of right-handed neutrinos has to be $M_N = \mathcal{O}(10^{14}-10^{15})\,\text{GeV}$ for $Y_\nu = \mathcal{O}(1)$. This is close to the scale of gauge unification $M_{\text{GUT}} = 10^{16}\,\text{GeV}$ in supersymmetric models. As we will see, the seesaw mechanism can be implemented into Grand Unified Theories. Especially in SO(10) models, the existence of right-handed neutrinos is naturally motivated, and the magnitude of their mass can be associated with the SO(10)-breaking scale or an intermediate SU(5) scale.

The Majorana nature of neutrinos implies two additional complex phases in the Lagrangean, which cannot be absorbed by a redefinition of the fermion fields. They enter the lepton mixing matrix V_{PMNS}, which therefore contains three physical phases.

1.2 Supersymmetric flavour physics

The extension of the Standard Model by supersymmetry (SUSY) stabilizes the scale of electroweak symmetry breaking, as was explained in the Introduction. The Standard Model is supersymmetrized by extending its particle content in order to assign each fermion a bosonic superpartner and vice versa. This extension is called the Minimal Supersymmetric Standard Model (MSSM).[6] Particles and their superpartners come in supermultiplets of equal fermionic and bosonic degrees of freedom. Chiral supermultiplets comprise the Standard-Model fermions together with their scalar superpartners, the sfermions, and two scalar Higgs boson doublets H_u, H_d joined with fermionic higgsinos, see Tab. 1.3. Vector supermultiplets include the gauge bosons of the Standard Model and assigned Majorana fermions of spin 1/2, the gauginos (Tab. 1.4). Note that all superpartners transform under the gauge group of the Standard Model as the corresponding particles, see Tabs. 1.1, 1.2. After electroweak symmetry breaking, charged higgsinos $\widetilde{H}_u^+$, $\widetilde{H}_d^-$ mix with winos $\widetilde{W}^\pm = (\widetilde{W}^1 \mp i\widetilde{W}^2)/\sqrt{2}$ to form two mass eigenstates, the charginos $\widetilde{\chi}_{1,2}^\pm$. Similarly, the mixing of the neutral higgsinos $\widetilde{H}_u^0$, $\widetilde{H}_d^0$ with the wino $\widetilde{W}^3$ and bino $\widetilde{B}$ gives rise to neutralinos $\widetilde{\chi}_{1,2,3,4}^0$.

---

[6]Properly defined, the MSSM is the minimal extension of the Standard-Model gauge group by an $N=1$ supersymmetry with R parity.

	quarks			leptons		higgsinos	
spin 1/2	$Q=(u_L,d_L)$	u_R	d_R	$L=(\nu_L,e_L)$	e_R	$(\widetilde{H}_u^+,\widetilde{H}_u^0)$	$(\widetilde{H}_d^0,\widetilde{H}_d^-)$
spin 0	$\widetilde{Q}=(\widetilde{u}_L,\widetilde{d}_L)$	$\widetilde{u}_R$	$\widetilde{d}_R$	$\widetilde{L}=(\widetilde{\nu}_L,\widetilde{e}_L)$	$\widetilde{e}_R$	(H_u^+,H_u^0)	(H_d^0,H_d^-)
	squarks			sleptons		Higgs bosons	

Table 1.3: Chiral supermultiplets in the MSSM.

	gluons	W bosons	B boson
spin 1	g^a	W^a	B
spin 1/2	$\widetilde{g}^a$	$\widetilde{W}^a$	$\widetilde{B}$
	gluinos	winos	bino

Table 1.4: Vector supermultiplets in the MSSM.

Two-Higgs-doublet model and soft supersymmetry breaking

The structure of supersymmetric Yukawa interactions is prescribed by the superpotential

$$W_{\text{MSSM}} = Y_u^{ij}\, Q_i u_j^c H_u - Y_d^{ij}\, Q_i d_j^c H_d - Y_e^{ij}\, L_i e_j^c H_d + \mu\, H_u H_d\,, \quad (1.20)$$

where $i,j = 1,2,3$ are flavour indices. The higgsino mass parameter μ is a complex quantity of mass dimension one. SU(2) doublets are contracted in a gauge-invariant way as in Eq. (1.5). The fields occurring in the superpotential are chiral superfields, cf. Tab. 1.3. Thus, fermions and sfermions couple with the same strength to either of the Higgs doublets

$$H_u = \begin{pmatrix} H_u^+ \\ H_u^0 \end{pmatrix} \quad \text{and} \quad H_d = \begin{pmatrix} H_d^0 \\ H_d^- \end{pmatrix}, \quad (1.21)$$

that transform under G_{SM} as $H_u : (1,2)_1$ and $H_d : (1,2)_{-1}$, respectively. We observe that, contrary to the Standard Model, the Higgs sector of the MSSM is a two-Higgs-doublet model (2HDM)[7] of type II. Since supersymmetry requires that the superpotential be a holomorphic function of chiral superfields (i.e., the simultaneous occurrence of H and H^* is not possible), two Higgs doublets are needed to separately give masses to both up- and down-type fermions. Below the scale of electroweak symmetry breaking, the neutral Higgs components exhibit vacuum expectation values $\langle H_u^0 \rangle = v_u$, $\langle H_d^0 \rangle = v_d$. These vevs are related to the vev of the Standard-Model Higgs boson via

$$v^2 = v_u^2 + v_d^2\,, \qquad \tan\beta = v_u/v_d; \quad (1.22)$$

their ratio $\tan\beta$ is a free parameter in the MSSM. This Higgs mechanism generates the fermion masses

$$m_u = y_u \cdot v_u = y_u \cdot v \sin\beta\,, \qquad m_{d,e} = y_{d,e} \cdot v_d = y_{d,e} \cdot v \cos\beta\,. \quad (1.23)$$

---
[7]For an extensive introduction into the Two-Higgs-Doublet model, see Ref. [19].

1.2. Supersymmetric flavour physics

In the course of electroweak symmetry breaking, three of the eight degrees of freedom in H_u and H_d, denoted by G^0 and $G^\pm$, give rise to longitudinal modes of the gauge bosons Z and $W^\pm$. The remaining degrees of freedom become manifest in five physical Higgs bosons: Two neutral scalars h^0 and H^0, one neutral pseudoscalar A^0, and two charged fields H^+ and H^-. By writing down the Higgs potential in the 2HDM and diagonalizing the resulting Higgs mass matrix, the Higgs mass eigenstates arise from the initial Higgs-doublet components as

$$\sqrt{2}\begin{pmatrix}\mathrm{Im}H_d^0\\ \mathrm{Im}H_u^0\end{pmatrix} = \begin{pmatrix}\sin\beta & -\cos\beta\\ \cos\beta & \sin\beta\end{pmatrix}\begin{pmatrix}A^0\\ G^0\end{pmatrix},\quad \sqrt{2}\begin{pmatrix}\mathrm{Re}H_d^0 - v_d\\ \mathrm{Re}H_u^0 - v_u\end{pmatrix} = \begin{pmatrix}\cos\alpha & -\sin\alpha\\ \sin\alpha & \cos\alpha\end{pmatrix}\begin{pmatrix}H^0\\ h^0\end{pmatrix},$$

$$\begin{pmatrix}H_d^{-*}\\ H_u^+\end{pmatrix} = \begin{pmatrix}\sin\beta & -\cos\beta\\ \cos\beta & \sin\beta\end{pmatrix}\begin{pmatrix}H^+\\ G^+\end{pmatrix},\quad \tan(2\alpha) = \tan(2\beta)\frac{M_{A^0}^2 + M_Z^2}{M_{A^0}^2 - M_Z^2},$$
(1.24)

with the Higgs mixing angles α and β. The corresponding Higgs boson masses at tree level are related by

$$M_{H^0,h^0}^2 = \frac{1}{2}\left\{M_{A^0}^2 + M_Z^2 \pm \sqrt{(M_{A^0}^2 + M_Z^2)^2 - 4M_{A^0}^2 M_Z^2 \cos^2(2\beta)}\right\},$$

$$M_{H^\pm}^2 = M_{A^0}^2 + M_W^2,$$
(1.25)

so that the spectrum of the 2HDM is determined by two parameters, which are customarily chosen to be $(\tan\beta, M_{A^0})$ or equivalently $(\tan\beta, M_{H^+})$. The mixing angles α and β enter the Higgs-fermion couplings and lead to interesting flavour effects in FCNC and charged currents, especially if $\tan\beta$ is large. Such effects will be the subject of Chapter 4.

From the vacuum state of the superpotential, we read off that the masses of particles and their superpartners have to be equal. Since, however, any direct searches of light scalar particles failed, one concludes that supersymmetry is not exact in nature, but has to be broken. In order to generate higher masses for superpartners and at the same time maintain the solution of the hierarchy problem, SUSY breaking has to be explicit, but "soft". The mechanism of SUSY breaking being unknown, we assume its origin at a high energy scale and parametrize its effects by softly SUSY-breaking terms added to the Lagrangean,

$$\begin{aligned}\mathcal{L}_{\mathrm{soft}} = &-\widetilde{Q}^* M_{\widetilde{Q}}^2 \widetilde{Q} - \widetilde{u}^c M_{\widetilde{u}}^2 \widetilde{u}^{c*} - \widetilde{d}^c M_{\widetilde{d}}^2 \widetilde{d}^{c*} - \widetilde{L}^* M_{\widetilde{L}}^2 \widetilde{L} - \widetilde{e}^c M_{\widetilde{e}}^2 \widetilde{e}^{c*}\\ &-\widetilde{Q}\,A_u\,\widetilde{u}^c H_u + \widetilde{Q}\,A_d\,\widetilde{d}^c H_d + \widetilde{L}\,A_e\,\widetilde{e}^c H_d + \mathrm{h.c.}\\ &-\frac{1}{2}(M_1\,\widetilde{B}\widetilde{B} + M_2\,\widetilde{W}^a\widetilde{W}^a + M_3\,\widetilde{g}^a\widetilde{g}^a) + \mathrm{h.c.}\\ &-m_{H_u}^2 H_u^* H_u - m_{H_d}^2 H_d^* H_d - (m_{12}^2 H_u H_d + \mathrm{h.c.})\,.\end{aligned}$$
(1.26)

Here, the sfermion masses M_f^2 are hermitean 3×3 matrices in flavour space. The complex trilinear couplings A_f of mass dimension one are analogous to the (dimensionless) Yukawa couplings in the superpotential in Eq. (1.20). Further, $\mathcal{L}_{\mathrm{soft}}$ comprises the gaugino terms with masses $M_{1,2,3}$ and SUSY-breaking contributions to the Higgs potential, parametrized by $m_{H_u}^2$, $m_{H_d}^2$, and m_{12}^2. In order to keep corrections to the light Higgs mass moderate, the mass scale associated with the soft parameters should not exceed $M_{\rm SUSY} \simeq 1\,\rm TeV$. By introducing these mass terms and trilinear couplings, we add a large number of free parameters that strongly reduce the predictivity of our model. In particular, the complex entries of M_f^2 and A_f are new sources of flavour and CP violation in addition to the CKM matrix. Facing experiment, however, the flavour structure of soft terms is rigorously constrained by K and B physics observables [20–22]. This motivates the assumption of minimal flavour violation (MFV), where "all flavour- and CP-violating interactions are linked to the known structure of Yukawa couplings" [23]. Within the framework of MFV, departures from Standard-Model flavour mixing due to new physics can be accurately distinguished.

Flavour mixing from soft mass splitting

To make the above statement explicit, let us cast flavour violation into formulae: In the fermion mass eigenbasis, the Yukawa terms in the superpotential read

$$W_{\rm MSSM}^Y = u\,\widehat{Y}_u\,u^c\,H_u^0 - d\,V_{\rm CKM}^\top\,\widehat{Y}_u\,u^c\,H_u^+ + d\,\widehat{Y}_d\,d^c\,H_d^0 - u\,V_{\rm CKM}^*\,\widehat{Y}_d\,d^c\,H_d^-$$
$$+ e\,\widehat{Y}_e\,e^c\,H_d^0 - \nu\,V_{\rm PMNS}^\top\,\widehat{Y}_e\,e^c\,H_d^-. \quad (1.27)$$

We observe that flavour mixing in charged-Higgs currents is governed by the matrices $V_{\rm CKM} = L_u^\dagger L_d$ and $V_{\rm PMNS} = L_e^\dagger L_\nu$, which also determine the flavour mixing in left-handed weak charged currents, cf. Eq. (1.10). The rotations of right-handed fermions are not visible in charged-Higgs currents (and in the supersymmetrized version involving fermions, sfermions, and higgsinos). Consequently, the supersymmetric part of the MSSM is minimally flavour-violating as the Standard Model. However, flavour violation in both left- and right-handed fermion sectors enters via SUSY breaking: The soft mass and trilinear A terms in Eq. (1.26) are not necessarily aligned with the Yukawa couplings. Now, minimal flavour violation is realized in a supersymmetric model if one assumes flavour-blind SUSY breaking at some scale $M \gtrsim M_{\rm SUSY}$, in our case at the Planck scale. Concretely, all soft mass parameters are universal, and trilinear couplings are proportional to the corresponding Yukawa couplings,

$$M_{\rm Pl}: \quad M_{\widetilde{Q}}^2 = M_{\widetilde{u}}^2 = M_{\widetilde{d}}^2 = M_{\widetilde{L}}^2 = M_{\widetilde{e}}^2 = m_0^2 \cdot \mathbf{1},$$
$$A_u = a_0 \cdot Y_u, \quad A_d = a_0 \cdot Y_d, \quad A_e = a_0 \cdot Y_e. \quad (1.28)$$

The masses and couplings in a renormalizable quantum field theory are energy-dependent. The mathematical framework of renormalization group evolution (RGE) links the values of those parameters at different scales. By evolving the soft parameters from the high scale $M_{\rm Pl}$ down to the electroweak scale M_Z, the universality in Eq. (1.28) is broken.

In particular, effects of the large top Yukawa coupling in the RGE separate the third-generation $\tilde{u}_L$, $\tilde{d}_L$, and $\tilde{u}_R$ squark masses from the light generations,

$$M_Z: \quad M_{\tilde{Q}}^2 = m_{\tilde{Q}}^2 \cdot \begin{pmatrix} 1 & 0 & 0 \\ 0 & 1 & 0 \\ 0 & 0 & 1-\Delta_{\tilde{Q}} \end{pmatrix}, \quad M_{\tilde{u}}^2 = m_{\tilde{u}}^2 \cdot \begin{pmatrix} 1 & 0 & 0 \\ 0 & 1 & 0 \\ 0 & 0 & 1-\Delta_{\tilde{u}} \end{pmatrix}, \quad (1.29)$$

$$M_{\tilde{d}}^2 = m_{\tilde{d}}^2 \cdot \mathbf{1}, \quad M_{\tilde{L}}^2 = m_{\tilde{L}}^2 \cdot \mathbf{1}, \quad M_{\tilde{e}}^2 = m_{\tilde{e}}^2 \cdot \mathbf{1}\,.$$

The universality of right-handed down-squark and lepton masses is preserved due to the smallness of bottom and tau Yukawa couplings. The flavour effects of the mass splitting become visible in the super-CKM basis, which is obtained by performing simultaneous rotations on quark and squark fields when diagonalizing the quark mass matrices. So, in the super-CKM basis, the mass splitting in $M_{\tilde{Q}}^2$ leads to off-diagonal elements in the mass matrix of left-handed down squarks,

$$(M_{\tilde{Q}}^2)^{\text{sCKM}} = V_{\text{CKM}}^\dagger M_{\tilde{Q}}^2 V_{\text{CKM}}\,. \quad (1.30)$$

In summary, in the MSSM with minimal flavour violation, there arise flavour-changing neutral currents of left-handed down squarks, which are guided by CKM mixing. Flavour mixing among sleptons and right-handed squarks is absent, so that one still cannot probe the rotation matrices R_f defined in Eq. (1.7).

1.3 Supersymmetric Yukawa unification

The prediction of gauge coupling unification is probably the most attractive feature of Grand Unified Theories. In non-supersymmetric GUTs, however, the measured gauge couplings do not meet when evolved to high energies. Supersymmetry helps to solve this problem. Namely, in the MSSM with superpartners at $M_{\text{SUSY}} \simeq 1\,\text{TeV}$, the three gauge couplings of the Standard Model converge at the scale $M_{\text{GUT}} \simeq 10^{16}\,\text{GeV}$ due to the effect of supersymmetric particles in the RGE [24–26]. Embedding the MSSM into a Grand Unified Theory extends Standard-Model flavour mixing by relating the (s)quark and (s)lepton sectors. Due to the unification of fermions in larger multiplets, the top Yukawa coupling now enters the RGE of soft mass matrices other than $M_{\tilde{Q}}$ and $M_{\tilde{u}}$ above the GUT scale. In SO(10), *all* soft mass matrices exhibit a mass splitting $\Delta_{\tilde{f}}$ for the third generation. This leads to FCNC of right-handed squarks and sleptons, and the rotation matrices R_f become physical. We will start by focussing on the implications of supersymmetric Grand Unification for flavour mixing in SUSY SU(5) as an intermediate step to explain supersymmetric SO(10) unification. Mathematical details are treated in the following sections.

SU(5)

SU(5) is the minimal enlargement of the Standard-Model gauge group G_{SM} in which all three gauge couplings $g_1' = \sqrt{5/3}\,g_1$, g_2, and g_3 unify to one coupling g_5 [27].[8] Even though G_{SM} is a subgroup of SU(5), it is not obvious that one can embed the fermions of the Standard Model into SU(5) representations without introducing new particles. Fortunately, the Standard-Model fermions (together with their corresponding superpartners) fit into the representations $\bar{5}$ and 10 of SU(5), in matrix notation

$$\bar{5} = \begin{pmatrix} d_1^c \\ d_2^c \\ d_3^c \\ e \\ -\nu_e \end{pmatrix}_L , \quad 10 = \frac{1}{\sqrt{2}} \begin{pmatrix} 0 & u_3^c & -u_2^c & -u_1 & -d_1 \\ -u_3^c & 0 & u_1^c & -u_2 & -d_2 \\ u_2^c & -u_1^c & 0 & -u_3 & -d_3 \\ u_1 & u_2 & u_3 & 0 & -e^c \\ d_1 & d_2 & d_3 & e^c & 0 \end{pmatrix}_L . \qquad (1.31)$$

All particles are left-handed Weyl spinors, and indices $1,2,3$ indicate colour, the quantum number of $SU(3)_C$. To generate neutrino masses via the seesaw mechanism, one adds an SU(5) singlet 1 for the heavy right-handed neutrino N with Yukawa coupling Y_N. The Yukawa terms in the superpotential are composed as SU(5)-invariant trilinears,

$$W_{\text{SU}(5)}^Y = \frac{1}{4} Y_5^{ij}\, 10_i 10_j 5_H + \sqrt{2}\, Y_{\bar{5}}^{ij}\, 10_i \bar{5}_j \bar{5}_H + Y_\nu^{ij}\, \bar{5}_i 1_j 5_H + \frac{1}{2} M_N^{ij}\, 1_i 1_j . \qquad (1.32)$$

The Higgs fields in SU(5) decompose into Standard-Model representations as

$$5_H = (H_C, -H_u), \quad \bar{5}_H = (\overline{H}_C, \overline{H}_d) . \qquad (1.33)$$

Besides the two MSSM Higgs doublets H_u and $(\overline{H}_d)^\alpha = \epsilon^{\alpha\beta}(H_d)_\beta$, cf. Eq. (1.21), they contain heavy Higgs fields transforming under G_{SM} as colour triplets, $H_C : (3,1)_0$ and $\overline{H}_C : (\bar{3},1)_0$. These coloured Higgs bosons and the heavy neutrinos are integrated out at and little below the SU(5) scale M_{GUT}, respectively, such that they do not appear as degrees of freedom in the low-energy Lagrangean. By calculating the Yukawa interactions in terms of MSSM fields, one identifies relations between the Yukawa couplings,

$$Y_u = (Y_5)_S \quad \text{and} \quad Y_d = Y_e^\top = Y_{\bar{5}} . \qquad (1.34)$$

The up-quark Yukawa coupling is thus restricted by SU(5) to be symmetric (denoted by $(Y_5)_S = (Y_5 + Y_5^\top)/2$). The Yukawa couplings of down quarks and charged leptons are unified, since these are embedded into the same SU(5) multiplet. This implies that the mixings of right-handed (left-handed) down quarks and left-handed (right-handed) charged leptons (cf. Eq. (1.7)) are identical up to complex conjugation,

$$R_d^* = L_e, \quad L_d = R_e^* . \qquad (1.35)$$

---

[8]The factor $\sqrt{5/3}$ in the $U(1)$ coupling is due to the fact that the properly normalized hypercharge generator in SU(5) is $\sqrt{3/5}\,Y$.

1.3. Supersymmetric Yukawa unification

In the U basis, where Y_u is diagonal, the Yukawa sector in terms of MSSM mass eigenstates is given by

$$W_{\text{MSSM}}^U = Q\widehat{Y}_u u^c H_u - Q(V_q^* \widehat{Y}_d V_\ell) d^c H_d - e^c(V_q^* \widehat{Y}_e V_\ell) L H_d - L Y_\nu N H_u + \frac{1}{2} M_N NN. \tag{1.36}$$

The couplings Y_d and $Y_e^\top$ are diagonalized by one and the same bi-unitary transformation, $Y_d = V_q^* \widehat{Y}_{d,e} V_\ell = Y_e^\top$. We directly identify $V_q = L_u^\dagger L_d = V_{\text{CKM}}$. If the U basis is also the basis of right-handed neutrino mass eigenstates, we further have $V_\ell = L_e^\dagger L_\nu = V_{\text{PMNS}}$. We introduce the phase matrices

$$\Theta_L = \text{diag}(e^{-i\alpha_1}, e^{-i\alpha_4}, e^{-i\alpha_5}), \qquad \Theta_R = \text{diag}(1, e^{i(\alpha_1-\alpha_2)}, e^{i(\alpha_1-\alpha_3)}), \tag{1.37}$$

such that we can define V_{CKM} and V_{PMNS} in their standard parametrization with one CP-violating phase δ each, given in Eq. (1.9). Five of six phases in V_q can be absorbed by a redefinition of the quark fields, as in the Standard Model. Due to the unification of quarks and leptons, the six phases in V_{PMNS} are then physical. With V_{PMNS} in its standard parametrization we thus have $V_\ell = \Theta_L^\dagger V_{\text{PMNS}} \Theta_R^\dagger$. The additional phases are important sources of new CP-violating effects in flavour-changing neutral currents. For the neutrino angle $\theta_{13} \neq 0$, the matrix that rotates (s)leptons and right-handed down (s)quarks is given by,

$$V_\ell = \begin{pmatrix} \sqrt{\frac{2}{3}} c_{13} e^{i\alpha_1} & \frac{1}{\sqrt{3}} c_{13} e^{i\alpha_2} & s_{13} e^{i(\delta+\alpha_3)} \\ e^{i\alpha_4}\left(-\frac{1}{\sqrt{6}} - \frac{1}{\sqrt{3}} s_{13} e^{-i\delta}\right) & e^{i(-\alpha_1+\alpha_2+\alpha_4)}\left(\frac{1}{\sqrt{3}} - \frac{1}{\sqrt{6}} s_{13} e^{-i\delta}\right) & \frac{1}{\sqrt{2}} c_{13} e^{i(-\alpha_1+\alpha_3+\alpha_4)} \\ e^{i\alpha_5}\left(\frac{1}{\sqrt{6}} - \frac{1}{\sqrt{3}} s_{13} e^{-i\delta}\right) & e^{i(-\alpha_1+\alpha_2+\alpha_5)}\left(-\frac{1}{\sqrt{3}} - \frac{1}{\sqrt{6}} s_{13} e^{-i\delta}\right) & \frac{1}{\sqrt{2}} c_{13} e^{i(-\alpha_1+\alpha_3+\alpha_5)} \end{pmatrix}. \tag{1.38}$$

In the following, we will assume $\theta_{13} = 0$, in which case the standard phase δ disappears from the mixing matrix V_ℓ.

Since the up quarks are not unified with down quarks and charged leptons, the top Yukawa coupling does not affect the RGE of $M_{\tilde{d}}^2$ and $M_{\tilde{L}}^2$, analogously to the MSSM, cf. Eq. (1.29). Hence, provided that soft masses are universal at a high scale, the unification of right-handed down-(s)quark and left-handed (s)lepton mixings (manifest in *one* rotation matrix V_ℓ) is not observable in sfermion currents at tree level. However, fermion couplings to heavy right-handed neutrinos can change the flavour structure of sfermion soft mass matrices [28, 29]. In particular, the coupling of down (s)quarks to right-handed neutrinos and coloured Higgs(ino) fields involves V_ℓ via $(R_d^\dagger = L_e^\top)$[9]

$$d^c Y_\nu N H_C = d^c R_d^\dagger (L_\nu^* \widehat{Y}_\nu R_\nu^\top) R_\nu^* N H_C = d^c V_{\text{PMNS}}^* \widehat{Y}_\nu N H_C. \tag{1.39}$$

---
[9]Here we omit the phase difference between V_ℓ and V_{PMNS} parametrized by $\Theta_{L,R}$ in Eq. (1.37).

These terms introduce V_PMNS into the RGE of the right-handed down-squark mass matrix in SU(5) above the GUT scale. The large mixing angles in V_PMNS induce off-diagonal elements

$$(M_{\tilde d}^2)_{ij}^\text{sCKM} \simeq -\frac{1}{8\pi^2}(V_\text{PMNS}^*)_{ik}\, y_{\nu_k}^2 (V_\text{PMNS}^\top)_{kj} \cdot (3\,m_{\tilde d}^2 + |A_d|^2) \cdot \log \frac{M_\text{Pl}}{M_\text{GUT}} \qquad (i \neq j). \quad (1.40)$$

If the neutrino Yukawa couplings y_{ν_k} are sizeable, this generates FCNC among right-handed down squarks, observable in $b_R - s_R$, $b_R - d_R$, and $s_R - d_R$ transitions [30–33].

SO(10)

In SO(10), all fifteen Standard-Model fermions (and superpartners) of one generation are unified in one 16-dimensional spinor representation [34, 35],

$$16 = 1 \oplus 10 \oplus \bar 5 = (N,\, (Q,\, u^c,\, e^c),\, (d^c,\, L)). \quad (1.41)$$

This framework includes the right-handed neutrino, thus providing light neutrino masses via the seesaw mechanism in a natural way. Compared to SU(5), SO(10) incorporates not only gauge unification but in addition the complete unification of the fermions of each generation. The unified fermion couplings exhibit a universal flavour structure above the SO(10) scale M_{10}, which leaves significant imprints in low-energy flavour physics. Concretely, the large top Yukawa coupling $y_t = (Y_{10})_{33}$ now affects the RGE of *all* soft mass matrices, such that the mass of all third-generation sfermions separates from the (degenerate) light sfermions. In the U basis, the mass eigenbasis of up-type quarks, one has

$$M_\text{Pl}:\ M_{\tilde f}^2 = m_0^2 \cdot \mathbf{1} \quad \xrightarrow{\text{RGE}} \quad M_{10}:\ M_{\tilde f}^2 = m_{\tilde f}^2(M_{10}) \cdot \begin{pmatrix} 1 & 0 & 0 \\ 0 & 1 & 0 \\ 0 & 0 & 1-\Delta_{\tilde f} \end{pmatrix}. \quad (1.42)$$

In particular, the soft mass matrix of right-handed down squarks at the electroweak scale is now given by

$$M_Z:\ (M_{\tilde d}^2)^U = m_{\tilde d}^2 \cdot \text{diag}(1, 1, 1 - \Delta_{\tilde d}), \quad (1.43)$$

with the mass splitting $\Delta_{\tilde d}$ found as [36]

$$m_{\tilde d}^2 \Delta_{\tilde d} = \frac{1}{8\pi^2}\, y_t^2 \cdot (3\,m_{\tilde d}^2 + |A_d|^2) \cdot (5\log\frac{M_\text{Pl}}{M_{10}} + \log\frac{M_{10}}{M_\text{GUT}}). \quad (1.44)$$

Due to the large matter content of SO(10), the RGE of the soft mass matrices above M_{10} is enhanced by a factor 5 with respect to the running below the SO(10) scale. Therefore $\Delta_{\tilde d}$ is numerically of $\mathcal{O}(1)$ and about three times larger than the corresponding SU(5) effect for $y_{\nu_3} = y_t$ in Eq. (1.40). We choose $M_{10} = 10^{17}$ GeV to be just one order of magnitude above M_GUT in order to maximize the mass splitting $\Delta_{\tilde d}$ from SO(10) running. Then the short range of SU(5) running has no significant further impact on the soft mass splitting.

1.3. Supersymmetric Yukawa unification

We further assume that the unified up-quark and light-neutrino Yukawa couplings are simultaneously diagonal with the right-handed neutrinos and separated from the unified down-quark and lepton Yukawa couplings. Thereby we are back in the flavour setup of the SU(5) scenario discussed before, apart from the SO(10)-specific large soft mass splitting $\Delta_{\tilde{d}}$ and the additional unification of Y_u and Y_ν. We rotate the down quarks into their mass eigenbasis (thereby reaching the super-CKM basis) via $V_\ell = \Theta_L^\dagger V_{\text{PMNS}} \Theta_R^\dagger$, cf. Eq. (1.36). Then $M_{\tilde{d}}^2$ is no longer diagonal, but exhibits large off-diagonal elements $(M_{\tilde{d}}^2)_{23}$ due to the large atmospheric neutrino mixing angle in V_ℓ,

$$(M_{\tilde{d}}^2)^{\text{sCKM}} = V_\ell^* (M_{\tilde{d}}^2)^U V_\ell^\top = m_{\tilde{d}}^2 \cdot \begin{pmatrix} 1 & 0 & 0 \\ 0 & 1 - \frac{\Delta_{\tilde{d}}}{2} & -\frac{\Delta_{\tilde{d}}}{2} e^{-i(\alpha_4-\alpha_5)} \\ 0 & -\frac{\Delta_{\tilde{d}}}{2} e^{i(\alpha_4-\alpha_5)} & 1 - \frac{\Delta_{\tilde{d}}}{2} \end{pmatrix}. \quad (1.45)$$

This observation has significant impact on $b-s$ transitions like $B_s - \bar{B}_s$ mixing [37] and $B \to X_s\gamma$ [38]. Moreover, the complex phase in $(M_{\tilde{d}}^2)_{23}^{\text{sCKM}}$ induces effects in CP-violating observables like the phase ϕ_s in $B_s - \bar{B}_s$ mixing and the CP asymmetries in $B_d \to \phi K_s$ or $B_s \to D_s K^\pm$ [36, 38].

1.4 Embedding the Standard Model into SO(10)

In the remaining sections of this chapter we discuss relevant mathematical aspects of SO(10) unification, still focussing on the Yukawa sector. Since unified fermions come with the same helicity, we use left-handed fields only. Right-handed fermions are embedded as charge-conjugated left-handed fields together with left-handed fermions into larger multiplets. These left-handed antiparticles transform as the conjugate representation $\bar{R}$, if the corresponding right-handed particles transform as R; for instance, $(u^c)_L$ transforms as $(\bar{3}, 1)_{-4/3}$, whereas u_R transforms as $(3, 1)_{4/3}$. Thus the overall $SU(3) \times SU(2) \times U(1)$ representation of left-handed fermion fields differs from the representation of right-handed fermions, because the subgroup $SU(2) \times U(1)$ is chiral. Consequently, the unifying representation has to be complex in order to allow $R \neq \bar{R}$. The group SO(10) provides such a complex representation, the 16-dimensional spinor representation. All fifteen Standard-Model fermions of one generation fit into a 16-plet of SO(10), in addition to a heavy right-handed neutrino N.

In order to understand how to embed the Standard-Model gauge group into SO(10), let us describe the embedding $SU(n) \subset SO(2n)$ in a general formalism [39, 40]. The group $SO(2n)$ is formed by $2n(2n-1)/2$ traceless antisymmetric objects $M_{\mu\nu}$ fulfilling the algebra

$$[M_{\mu\nu}, M_{\rho\sigma}] = -i(\delta_{\mu\sigma}M_{\nu\rho} + \delta_{\nu\rho}M_{\mu\sigma} - \delta_{\mu\rho}M_{\nu\sigma} - \delta_{\nu\sigma}M_{\mu\rho}), \quad (1.46)$$

with $\mu, \nu, \rho, \sigma = 1, \ldots, 2n$ being SO(10) indices. The fundamental representation is given by the antisymmetric $2n \times 2n$ matrices

$$(M_{\mu\nu})_{\alpha\beta} = -i(\delta_{\mu\alpha}\delta_{\nu\beta} - \delta_{\nu\alpha}\delta_{\mu\beta}). \quad (1.47)$$

The generators of the spinor representation of $SO(2n)$ are constructed from a set of hermitean matrices Γ_μ, which fulfill the Clifford algebra $\{\Gamma_\mu, \Gamma_\nu\} = 2\delta_{\mu\nu}$. The objects

$$\Sigma_{\mu\nu} = \tfrac{1}{4i}[\Gamma_\mu, \Gamma_\nu] \quad (1.48)$$

obey the commutation relations in Eq. (1.46), so these are the desired generators of the $SO(2n)$ spinor representation. The matrices Γ_μ can be constructed iteratively from the Pauli matrices,

$$\begin{aligned}
\Gamma_1^{(k=1)} &= \sigma^1, \; \Gamma_2^{(1)} = \sigma^2, \; \Gamma_3^{(1)} = \sigma^3: \\
\Gamma_\mu^{(k+1)} &= \Gamma_\mu^{(k)} \otimes \Gamma_3^{(1)}, \quad \mu = 1, \ldots, 2n, \\
\Gamma_{2k+1}^{(k+1)} &= \mathbf{1}^{(k)} \otimes \Gamma_1^{(1)}, \\
\Gamma_{2k+2}^{(k+1)} &= \mathbf{1}^{(k)} \otimes \Gamma_2^{(1)}.
\end{aligned} \quad (1.49)$$

From these $2n$ operators $\Gamma_\mu^{(n)}$ one can find a set of n creation and annihilation operators

$$\begin{aligned}
b_a^\dagger &= \tfrac{1}{2}(\Gamma_{2a-1} + i\Gamma_{2a}), \quad a = 1, \ldots, n, \\
b_a &= \tfrac{1}{2}(\Gamma_{2a-1} - i\Gamma_{2a}),
\end{aligned} \quad (1.50)$$

1.4. Embedding the Standard Model into SO(10)

which fulfill the anticommutation relations

$$\{b_a, b_b^\dagger\} = \delta_{ab}, \quad \{b_a, b_b\} = \{b_a^\dagger, b_b^\dagger\} = 0. \tag{1.51}$$

An SU(n) algebra is eventually generated by

$$T_{ab} = b_a^\dagger b_b : \quad [T_{ab}, T_{cd}] = \delta_{bc} T_{ad} - \delta_{ad} T_{cb}. \tag{1.52}$$

As the SU(n) generators T_{ab} can be expressed in terms of $\Sigma_{\mu\nu}$, this is indeed a subalgebra of SO($2n$). Note that the operators $\Gamma_\mu^{(k)}$ in Eq. (1.49) are direct products of k SU(2) generators $\Gamma_\mu^{(1)}$, such that one can think of an SO($2n$) spinor state as a sequence of n signs, $|\pm \pm \cdots \pm\rangle$. The creation and annihilation operators $b_a^\dagger \sim \Gamma_1^{(1)} + i\Gamma_2^{(1)}$ and $b_a \sim \Gamma_1^{(1)} - i\Gamma_2^{(1)}$ flip the sign at position a like $(- \to +)$ and $(+ \to -)$, respectively. This defines a vacuum state $|--\cdots-\rangle \equiv |0\rangle$ that is annihilated by all operators b_a. In the spinor representation of SO($2n$), the generators $\Sigma_{\mu\nu}$ act reducibly on a 2^n-dimensional space, which can be decomposed into two irreducible subspaces characterized by an even (odd) number of signs $\pm$.

Let us write down explicitly the embedding SU(n) $\subset$ SO($2n$) for $n = 5$: In the SU(5) basis, a general SO(10) spinor is given in one of the 16-dimensional irreducible representations by

$$\begin{aligned}
|16\rangle &= |0\rangle \psi_0 + \frac{1}{2} b_a^\dagger b_b^\dagger |0\rangle \psi^{ab} + \frac{1}{4!} \epsilon^{abcde} b_b^\dagger b_c^\dagger b_d^\dagger b_e^\dagger |0\rangle \bar{\psi}_a \quad \text{or} \\
|\overline{16}\rangle &= b_a^\dagger |0\rangle \psi^a + \frac{1}{2 \cdot 3!} \epsilon^{abcde} b_c^\dagger b_d^\dagger b_e^\dagger |0\rangle \bar{\psi}_{ab} + b_1^\dagger b_2^\dagger b_3^\dagger b_4^\dagger b_5^\dagger |0\rangle \bar{\psi}_0.
\end{aligned} \tag{1.53}$$

The objects ψ_0, ψ^{ab}, and $\bar{\psi}_a$ transform under SU(5) as the singlet, 10-dimensional, and conjugate 5-dimensional $\bar{5}$ representation. All indices are antisymmetrized, respecting the anticommutation properties of the operators $b_a^\dagger$ and b_a. In a last step we identify the Standard-Model SU(3) × SU(2) representations embedded in SU(5). By splitting the set of operators $b_a^{(\dagger)}$ into two subsets $(b_1^{(\dagger)}, b_2^{(\dagger)}, b_3^{(\dagger)})$ and $(b_4^{(\dagger)}, b_5^{(\dagger)})$, we create subspaces for SU(3) and SU(2). This decomposition allows us to identify the Standard-Model fermions as objects in the SU(5) basis, using the $|16\rangle$ representation in Eq. (1.53),

$$\begin{aligned}
\psi_0 : \quad & N = \quad |---;--\rangle \\
\psi^{ab} : \quad & Q = \quad \left(|+--;{}^{+-}_{-+}\rangle, |-+-;{}^{+-}_{-+}\rangle, |--+;{}^{+-}_{-+}\rangle\right) \\
& u^c = \quad \left(|++-;--\rangle, |+-+;--\rangle, |-++;--\rangle\right) \\
& e^c = \quad |---;++\rangle \\
\bar{\psi}_a : \quad & d^c = \quad \left(|++-;++\rangle, |+-+;++\rangle, |-++;++\rangle\right) \\
& L = \quad |+++;{}^{+-}_{-+}\rangle.
\end{aligned} \tag{1.54}$$

In summary, the fermion fields of the Standard Model, together with the right-handed neutrino, are embedded in a spinor representation of SO(10) via an intermediate SU(5) symmetry in the following way,

$$16 = 1 \oplus 10 \oplus \bar{5} = (\psi_0, \psi^{ab}, \bar{\psi}_a) = (N, (Q, u^c, e^c), (d^c, L)). \tag{1.55}$$

1.5 SO(10) breaking

The mechanism of breaking SO(10) down to the Standard-Model gauge group $G_{\rm SM} = {\rm SU}(3) \times {\rm SU}(2) \times {\rm U}(1)$ has large impact on the Higgs field content of a specific SO(10) model. Consequently, it will determine the structure of Yukawa interactions between matter and Higgs fields, which is our focus. Here we discuss two main roads of breaking SO(10) $\to G_{\rm SM}$ (see also e.g. Ref. [39]). The group SO(10) can be broken to the Standard Model via two subgroups, SU(5) and the so-called Pati-Salam group $G_{\rm PS} = {\rm SU}(4) \times {\rm SU}(2) \times {\rm SU}(2)$ [41]. Though direct breaking SO(10) $\to G_{\rm SM}$ is possible, models with intermediate symmetries are advantageous: The mass scale essential to implement the seesaw mechanism can be associated with one of these subgroups.

With the Pati-Salam group as an intermediate symmetry, the breaking mechanism incorporates the symmetric 54 and a 16 or 126 representation. The 54 (also 210) potentially exhibits a vacuum expectation value leaving the SO(10) subgroup SO(6)$\times$SO(4) untouched, which is isomorphic to $G_{\rm PS}$. The Pati-Salam group is then broken to the Standard-Model group $G_{\rm SM}$ using a 16 or 126 representation.

When opting for SU(5) as intermediate symmetry, a suitable representation to break SO(10) is the antisymmetric adjoint 45. Its decomposition into SU(5) representations reads

$$45 = 24 \oplus 10 \oplus \overline{10} \oplus 1. \tag{1.56}$$

A vev in the SU(5)-singlet direction, $\langle 1 \rangle_{45} \equiv v_{10} \sim M_{10}$, breaks SO(10) $\to$ SU(5) $\times$ U(1). The additional U(1) is broken by the spinor Higgs representation 16. This reduces the rank of the group from 5 for SO(10) to 4 for SU(5) and $G_{\rm SM}$. The subsequent breaking SU(5) $\to G_{\rm SM}$ can be achieved using the SU(5)-adjoint representation 24 in 45. A vev $\langle 24 \rangle_{45} \equiv v_5 \sim M_{\rm GUT}$ proportional to the hypercharge generator of SU(5),

$$\langle 24 \rangle_{45} \equiv v_5 \, {\rm diag}(2,2,2\,;-3,-3)\,, \tag{1.57}$$

has the desired properties for this breaking.
Finally, one has to arrange for electroweak symmetry breaking SU(3)$_C$ $\times$ SU(2)$_L$ $\times$ U(1)$_Y$ $\to$ SU(3)$_C$ $\times$ U(1)$_{\rm em}$. To give the fermions mass, one needs a representation that couples to the fermion bilinears and contains a component that transforms under $G_{\rm SM}$ as the Standard-Model Higgs field $(1,2)_1$. Both requirements are fulfilled by the 5 and 45 representations of SU(5), see Eq. (A.10). In the fermion bilinear $16 \otimes 16$ each of the three constituents 10, 120, and 126 contains a 5 and/or 45 representation and can therefore contribute to electroweak symmetry breaking. (The decompositions of SO(10) representations are given in Appendix A.2.)
The pattern of SO(10) breaking discussed here is summarized in Fig. 1.3. In Chapter 2, we will concentrate on the breaking chain

$$ {\rm SO}(10) \xrightarrow[16,\overline{16}]{45_{(1)}} {\rm SU}(5) \xrightarrow{45_{(24)}} {\rm SU}(3) \times {\rm SU}(2) \times {\rm U}(1) \xrightarrow{10} {\rm SU}(3) \times {\rm U}(1) \,. \tag{1.58}$$

1.6. Yukawa unification in SO(10)

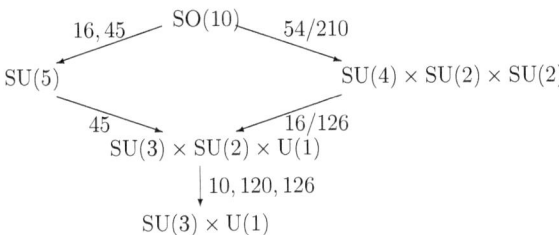

Figure 1.3: SO(10) breaking to the Standard Model. The Higgs representations noted with the arrows refer to SO(10).

The representation $\overline{16}$ is needed together with 16 in supersymmetric models in order to preserve supersymmetry when reducing the rank of the gauge group [42]. Without adding further representations, the possible content of Higgs fields in the Yukawa sector will consist of 10_H, 16_H, $\overline{16}_H$, and 45_H.

1.6 Yukawa unification in SO(10)

Using the tools introduced in Sec. 1.4, we proceed and construct SO(10)-invariant Yukawa terms. The tensor product of two 16-dimensional spinor representations decomposes into tensor representations like

$$16 \otimes 16 = 10_S \oplus 120_A \oplus 126_S, \qquad (1.59)$$

which have respective symmetric (S) and antisymmetric (A) structures. Consequently, there are three possibilities to form renormalizable Yukawa couplings of two fermion multiplets to a Higgs representation, namely the symmetric terms $16\,16\,10_H$ and $16\,16\,\overline{126}_H$, as well as an antisymmetric $16\,16\,120_H$ coupling. The resulting SO(10) superpotential reads

$$W^Y_{\text{ren}} = Y^{ij}_{10}\, 16_i\, 16_j\, 10_H + Y^{ij}_{120}\, 16_i\, 16_j\, 120_H + Y^{ij}_{126}\, 16_i\, 16_j\, \overline{126}_H\,. \qquad (1.60)$$

The indices $i,j = 1,2,3$ label the three generations of fermions. In the formalism of an SU(5) basis, the various terms are written as

$$\begin{aligned} 16_i\,16_j\,10_H :\quad & \langle 16^*_i | B\,\Gamma_\mu\,\phi_\mu | 16_j \rangle, \\ 16_i\,16_j\,120_H :\quad & \frac{1}{3!}\,\langle 16^*_i | B\,\Gamma_\mu\Gamma_\nu\Gamma_\rho\,\phi_{\mu\nu\rho} | 16_j \rangle, \\ 16_i\,16_j\,\overline{126}_H :\quad & \frac{1}{5!}\,\langle 16^*_i | B\,\Gamma_\mu\Gamma_\nu\Gamma_\rho\Gamma_\sigma\Gamma_\tau\,\phi_{\mu\nu\rho\sigma\tau} | 16_j \rangle. \end{aligned} \qquad (1.61)$$

The objects ϕ_μ, $\phi_{\mu\nu\rho}$, and $\phi_{\mu\nu\rho\sigma\tau}$ stand for the 10-, 120-, and 252-dimensional Higgs representations, the latter decomposing into $126 \oplus \overline{126}$ with only the $\overline{126}$ coupling to the spinors.

B is the charge conjugation matrix in SO(10), analogous to the charge conjugation matrix C of the Lorentz group, which is dropped here. SO(10) invariance of the trilinears in Eq. (1.61) is ensured if one chooses $B \equiv - \prod_{\mu \text{ even}} \Gamma_\mu$, so that

$$B^{-1} \Gamma_\mu^\top B = -\Gamma_\mu. \tag{1.62}$$

To elaborate on the details of SO(10) Yukawa unification, we exemplarily work out the coupling $16_i \, 16_j \, 10_H$ in Eq. (1.61) in terms of its SU(5) constituents. Using Eq. (1.53) and the algebra in Eq. (1.51), one gets

$$\langle 16^* | B = -i\,\psi_0 \langle 0 | \, b_1 b_2 b_3 b_4 b_5 - i \tfrac{1}{12} \epsilon_{abcde} \psi^{ab} \langle 0 | \, b_c b_d b_e - i \, \overline{\psi}_a \langle 0 | \, b_a \,. \tag{1.63}$$

The tensor representation 10_H is decomposed into irreducible SU(5) representations as $10 = 5 \oplus \bar{5}$, which we express in terms of ϕ_μ,

$$\begin{aligned} 5_H : \quad & \phi_{c_a} = \phi_{2a-1} - i\phi_{2a}\,, \quad a = 1, \ldots, 5, \\ \bar{5}_H : \quad & \phi_{\bar{c}_a} = \phi_{2a-1} + i\phi_{2a}\,, \end{aligned} \tag{1.64}$$

and thereby have

$$\Gamma_\mu \phi_\mu = b_a \phi_{\bar{c}_a} + b_a^\dagger \phi_{c_a}\,. \tag{1.65}$$

This formalism can be extended to the larger tensor representations $\phi_{\mu\nu\rho}$ and $\phi_{\mu\nu\rho\sigma}$ [43].[10] With these tools at hand, one computes the SO(10) Yukawa term in its SU(5) decomposition (after rescaling the Higgs fields $\phi_{c_a} = \sqrt{2} H^a$, $\phi_{\bar{c}_a} = \sqrt{2} \overline{H}_a$ to obtain canonical kinetic terms),

$$\langle 16_i^* | B\, \Gamma_\mu \phi_\mu | 16_j \rangle = i\sqrt{2} \Big\{ - (\psi_0^i \, \overline{\psi}_a^j + \overline{\psi}_a^i \, \psi_0^j) \, H^a \tag{1.66}$$

$$+ (\psi_i^{ab} \, \overline{\psi}_a^j + \overline{\psi}_a^i \, \psi_j^{ab}) \, \overline{H}_b + \tfrac{1}{4} \epsilon_{abcde} \psi_i^{ab} \, \psi_j^{cd} \, H^e \Big\},$$

short: $Y_{10}^{ij} \, 16_i 16_j \, 10_H \equiv Y_{10}^{ij} \sqrt{2} \Big\{ - (1_i \, \bar{5}_j + \bar{5}_i \, 1_j) \, 5_H + (10_i \, \bar{5}_j + \bar{5}_i \, 10_j) \, \bar{5}_H + \tfrac{1}{4} 10_i 10_j \, 5_H \Big\}.$

We observe that the Yukawa coupling Y_{10} has a symmetric flavour structure.[11] The representations 5_H and $\bar{5}_H$ contain components that transform under $G_{\rm SM}$ as $(1,2)_1$ and $(1,2)_{-1}$. After electroweak symmetry breaking, they give masses to up quarks and neutrinos or down quarks and charged leptons, respectively. With only one Yukawa term $Y_{10}^{ij} \, 16_i 16_j \, 10_H$ generating masses for all fermions, there is no mixing between fermions of different generations. Furthermore, the Yukawa matrices of all fermions are unified at the GUT scale and they are symmetric,

$$\sqrt{2}\,(Y_{10})_S = Y_u = Y_\nu^\top / 2 = Y_d = Y_e^\top, \tag{1.67}$$

---

[10]Note the differing definition of $\phi_{c_a}, \phi_{\bar{c}_a}$ in Eq. (1.64), which we adapted to the definition of $b_a^\dagger, b_a$ in Eq. (1.50).

[11]The factor i is absorbed into the Yukawa coupling Y_{10}.

with $(Y_{10})_S^{ij} = (Y_{10}^{ij}+Y_{10}^{ji})/2$. Since the fermion mass spectrum measured at low energies can agree with this relation for the third generation only, one has to enlarge the Yukawa sector. There are two main roads to correct the Yukawa unification of the light generations: First, one could add the renormalizable couplings with 120_H and $\overline{126}_H$ representations given in Eq. (1.61), corresponding to antisymmetric and symmetric Yukawa structures, respectively. By doing so, it is possible to accommodate the measured fermion masses and mixings [44]. However, adding large representations causes difficulties with the perturbativity of the gauge coupling above the SO(10) scale [45]. The second possibility to generate a realistic Yukawa structure is to introduce non-renormalizable terms of higher mass dimension. They can be built by means of small Higgs representations, such that the couplings stay perturbative up to the Planck scale. Such higher-dimensional terms, being suppressed by powers of a high mass scale, will correct the relations between masses and mixings of light fermions, without affecting the successful unification in the third generation.

1.7 Higher-dimensional Yukawa terms

In this section, we will elaborate on how to explicitly construct Yukawa corrections of mass dimension five. Staying with the small Higgs representations 10_H, 16_H, $\overline{16}_H$, and 45_H, one can build the following SO(10)-invariant dimension-five terms,

$$16\,16\,16_H 16_H, \quad 16\,16\,\overline{16}_H \overline{16}_H, \quad 16\,16\,10_H 45_H. \tag{1.68}$$

The compositions with spinor Higgs fields are suitable to generate Majorana masses for right-handed neutrinos via a vev in the SU(5)-singlet component of 16_H, $\overline{16}_H$. The term $16\,16\,10_H\,45_H$ provides manifold patterns for fermion masses and mixings. There are four different ways to form an SO(10) invariant out of its constituents, summarized in Ref. [46],

$$(16\,16)_{10}\,(10_H\,45_H)_{10} = \langle 16^*|B\,\Gamma_\mu \phi_\nu \phi_{\mu\nu}|16\rangle \tag{1.69a}$$
$$(16\,16)_{120}\,(10_H\,45_H)_{120} = \langle 16^*|B\,\Gamma_\mu \Gamma_\nu \Gamma_\rho \phi_\mu \phi_{\nu\rho}|16\rangle \tag{1.69b}$$
$$(16\,10_H)_{16^*}\,(16\,45_H)_{16} = \langle 16^*|B\,\Gamma_\mu \phi_\mu \Sigma_{\nu\rho} \phi_{\nu\rho}|16\rangle \tag{1.69c}$$
$$(16\,10_H)_{144^*}\,(16\,45_H)_{144} = \langle 16^*|B\,\phi_\mu \Gamma_\nu \phi_{\mu\nu}|16\rangle - (1.69c)\,, \tag{1.69d}$$

with the 10_H and 45_H represented by ϕ_μ and $\phi_{\mu\nu}$. Consequently, one can produce four different Yukawa textures. The terms in Eqs. (1.69a) and (1.69b) lead to the same symmetric and antisymmetric structures as the renormalizable terms $16\,16\,10_H$ and $16\,16\,120_H$, since the relevant component of the reducible representation $(10_H \otimes 45_H)$ couples to the fermion bilinear $(16 \otimes 16)$ as an effective 10_H or 120_H.

The appearance of dimension-five Yukawa terms is natural when thinking of SO(10) as an effective theory below the Planck scale. In this setup, dimension-five terms arise as effective operators by integrating out heavy degrees of freedom X above the SO(10) scale. Therefore the terms in Eqs. (1.69a) – (1.69d) are suppressed by the mass M_X of these heavy particles. The Feynman diagram corresponding to the effective 10 coupling in Eq. (1.69a)

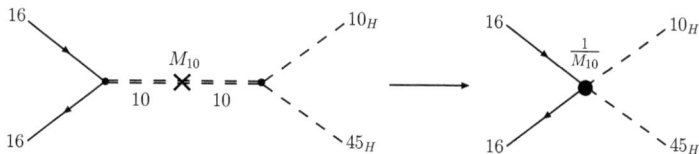

Figure 1.4: Non-renormalizable Yukawa term $(16\,16)_{10}\,(10_H\,45_H)_{10}$ of mass dimension five.

is illustrated in Fig. 1.4. The dashed double line denotes the heavy field 10, which is integrated out at the mass scale M_{10}.

Let us explicitly describe the formation of the term $(16\,16)_{10}\,(10_H\,45_H)_{10}$. By extending the formalism in Eq. (1.64) for higher tensor representations, one derives the decomposition of $\phi_{\mu\nu...}$ into irreducible SU(5) representations, in particular $\phi_{\mu\nu} \equiv 45_H = 1_{45} \oplus 10_{45} \oplus \overline{10}_{45} \oplus 24_{45}$. Details are given in Appendix A.2. Then one can express the effective coupling $(10_H\,45_H)_{10}$ in terms of SU(5) fields,

$$10\,10_H 45_H = \phi'_\mu \phi_\nu \phi_{\mu\nu} \tag{1.70}$$
$$= \tfrac{1}{\sqrt{2}}\left\{\tfrac{1}{\sqrt{5}}(5_{10}\bar{5}_H + \bar{5}_{10}5_H)1_{45} + 5_{10}\bar{5}_H\overline{10}_{45} + \bar{5}_{10}5_H 10_{45} + (5_{10}\bar{5}_H + \bar{5}_{10}5_H)24_{45}\right\}.$$

Since we want to break SU(5) such that the Standard-Model gauge symmetry $G_{\rm SM}$ is preserved, only the coupling to 24_{45} is relevant for Yukawa corrections, cf. Eq. (1.57). When integrating out the heavy field 10 by means of a mass term $2M_{10}\,5_{10}\bar{5}_{10}$, the coupling $16\,16\,10$ in Eq. (1.66) merges with the above $10\,10_H 45_H$, yielding the effective dimension-five Yukawa term

$$\langle 16_i^* | B\,\Gamma_\mu \phi_\nu \phi_{\mu\nu} | 16_j \rangle = \tfrac{i}{2}\left\{(\psi_i^{ab}\overline{\psi}_a^j + \overline{\psi}_a^i \psi_j^{ab})\Sigma_b^c \overline{H}_c + \tfrac{1}{4}\epsilon_{abcde}\psi_i^{ab}\psi_j^{cd}\Sigma_f^e H^f\right\},$$
$$\text{short:} \quad \frac{\widetilde{Y}_{10}^{ij}}{M_{10}} 16_i 16_j\,10_H 45_H \equiv \frac{\widetilde{Y}_{10}^{ij}}{M_{10}}\left\{(10_i\bar{5}_j + \bar{5}_i 10_j)24_{45}\bar{5}_H + \tfrac{1}{4}10_i 10_j\,24_{45}5_H\right\}. \tag{1.71}$$

Here we have suppressed the couplings to $(1_i\bar{5}_j + \bar{5}_i 1_j)$, which generate Dirac masses and Yukawa couplings of neutrinos. Explicit expressions of the dimension-five terms in Eqs. (1.69b) – (1.69d) are given in Ref. [46]. The complete Yukawa sector extended by dimension-five couplings reads

$$Y_{10}^{ij}\,16_i 16_j\,10_H + \frac{\widetilde{Y}_{10}^{ij}}{M_{10}}(16_i 16_j)_{10}(10_H 45_H)_{10} + \frac{\widetilde{Y}_{120}^{ij}}{M_{120}}(16_i 16_j)_{120}(10_H 45_H)_{120}$$
$$+ \frac{\widetilde{Y}_{16}^{ij}}{M_{16}}(16_i 16_j)_{16^*}(10_H 45_H)_{16} + \frac{\widetilde{Y}_{144}^{ij}}{M_{144}}(16_i 16_j)_{144^*}(10_H 45_H)_{144}. \tag{1.72}$$

The couplings Y_{10}^{ij} and $\widetilde{Y}_{10}^{ij}$ are flavour-symmetric, $\widetilde{Y}_{120}^{ij}$ and $\widetilde{Y}_{16}^{ij}$ are antisymmetric, and $\widetilde{Y}_{144}^{ij}$ is not restricted by SO(10) symmetry. One can thus generate corrections to the Yukawa

1.7. Higher-dimensional Yukawa terms

unification in Eq. (1.67) with arbitrary flavour structure. Setting the mass scales of all heavy fields equal to M_{Pl}, the Yukawa sector with non-renormalizable operators can be compressed to

$$W_{\text{nr}}^Y = Y_{10}^{ij}\, 16_i 16_j\, 10_H + \frac{\widetilde{Y}^{ij}}{M_{\text{Pl}}} 16_i 16_j\, 10'_H 45_H\,, \qquad (1.73)$$

with an arbitrary unitary *effective* Yukawa matrix $\widetilde{Y}$. Note that a second Higgs field $10'_H$ is needed in order to yield non-trivial fermion mixings. If the same representation 10_H appears in both dimension-four and dimension-five terms, the Yukawa couplings are factored out in the vacuum structure of W_{nr}^Y and the mixing matrix is unity. In case the dimension-five term has an effective 10 structure, the unification of down-type quarks and charged leptons is preserved, provided that they both couple exclusively to $10'_H$. Corrections for light fermions arise by adding effective 120, 16, and/or 144 terms. At the GUT scale, the SU(5)-breaking vev $\langle 24 \rangle_{45} = v_5 \operatorname{diag}(2,2,2;-3,-3)$ corrects down-quark-lepton unification by higher-dimensional Yukawa couplings, leading to

$$Y_d = Y_e^\top + 5\frac{v_5}{M_{\text{Pl}}}\widetilde{Y} \qquad \text{with}\quad \widetilde{Y} = \widetilde{Y}(\widetilde{Y}_{120},\widetilde{Y}_{16},\widetilde{Y}_{144})\,. \qquad (1.74)$$

The flavour structure of $\widetilde{Y}$, which is a priori arbitrary, can be constrained from flavour-changing processes involving light fermions. This will be extensively discussed in Chapter 3. As demonstrated in Sec. 1.3, a supersymmetric framework is needed to render Yukawa unification, and likewise its corrections, visible. We will show that FCNC constraints on $\widetilde{Y}$ have important consequences for SUSY GUT model building with higher-dimensional operators.

Chapter 2
A Supersymmetric SO(10) Model of Flavour

The unification of down quarks and charged leptons translates the large atmospheric neutrino mixing angle into supersymmetric $b - s$ transitions. These flavour-changing effects are particularly transparent in a SUSY SO(10) model developed by Chang, Masiero, and Murayama. After an introduction to the Yukawa sector of the model, we summarize flavour-independent constraints on the SUSY parameter space. We subsequently discuss the phenomenology of neutrino mixing in down-(s)quark currents, focussing on meson mixing observables. The chapter closes with a quantitative examination of unification effects in the mass difference and CP phase in $B_s - \bar{B}_s$ mixing.

2.1 The Chang-Masiero-Murayama model

In order to study the effects of large atmospheric neutrino mixing in $b - s$ transitions, in 2003, Chang, Masiero, and Murayama designed a supersymmetric perturbative SO(10) model for this purpose [36]. Their reasoning was the following: In an SO(10) fermion multiplet, the large top Yukawa coupling is unified with the third-generation neutrino Yukawa coupling. Due to the maximal atmospheric neutrino mixing $\theta_{23} \approx 45°$, the third-generation neutrino ν_3 consists in equal parts of ν_μ and ν_τ. The top quark is thus embedded in the multiplet together with the third-generation $\bar{5}$-plet of SU(5),

$$\bar{5}_3 = \bar{5}_\tau \cos\theta_{23} + \bar{5}_\mu \sin\theta_{23} \,. \tag{2.1}$$

The leptons τ and μ (with the corresponding neutrinos), in turn, are unified with right-handed down quarks,

$$\bar{5}_\tau = (b_1^c, b_2^c, b_3^c, \tau, -\nu_\tau), \qquad \bar{5}_\mu = (s_1^c, s_2^c, s_3^c, \mu, -\nu_\mu) \,. \tag{2.2}$$

Large $\nu_\tau - \nu_\mu$ mixing therefore implies large $b_R - s_R$ mixing, which is not observable in Standard-Model interactions due to the lack of right-handed flavour-changing currents. In a supersymmetric model, however, the large top Yukawa coupling generates a considerable mass correction of the third-generation down squark $\tilde{b}^c \cos\theta_{23} + \tilde{s}^c \sin\theta_{23}$, which induces large $\tilde{b}_R - \tilde{s}_R$ mixing. This summarizes our introduction to flavour mixing effects in SO(10) at the end of Sec. 1.3.

The Chang-Masiero-Murayama model (short: CMM model) is based on small SO(10) representations in order to ensure perturbative couplings up to the Planck scale. Consequently, the Yukawa sector of the superpotential is constructed by means of higher-dimensional operators,

$$W^Y_{\text{CMM}} = 16_i \, Y^{ij}_{10} \, 16_j \, 10_H + 16_i \, Y^{ij}_{45} \, 16_j \, \frac{45_H \, 10'_H}{M_{\text{Pl}}} + 16_i \, Y^{ij}_{16} \, 16_j \, \frac{\overline{16}_H \overline{16}_H}{2 M_{\text{Pl}}} \,. \tag{2.3}$$

The minimal content of Higgs fields in the model therefore is

$$10_H, \ 10'_H, \ 45_H, \ 16_H, \ \overline{16}_H \,. \tag{2.4}$$

Within this framework, SO(10) is broken to SU(5) by vevs $v_{16} \sim 10^{17}\,\text{GeV}$ in the pair of Higgs spinors 16_H and $\overline{16}_H$ and by $v_{10} \sim 10^{17}\,\text{GeV}$ in the SU(5)-singlet component of 45_H. Further, 45_H develops a smaller vev $v_5 \sim 10^{16}\,\text{GeV}$ in the SU(5)-adjoint component 24_H, cf. Eq. (1.57), which breaks SU(5) down to G_{SM}. Electroweak symmetry breaking is finally achieved via the vevs v_u and v_d of the Higgs doublets H_u and H_d contained in 10_H and $10'_H$. This is the breaking chain introduced already in Eq. (1.58). The CMM model is a "minimal perturbative SO(10) model" in the sense that the Yukawa sector is constructed out of Higgs representations needed for symmetry breaking anyway. Further, only the minimal set of dimension-five terms needed for realistic fermion masses and mixings is selected.[1] The three Yukawa terms in W^Y_{CMM} give masses to up quarks and Dirac neutrinos, down quarks and charged leptons, and right-handed neutrinos, respectively. In general, the MSSM Higgs doublets H_u and H_d are linear combinations of the light degrees of freedom in the 5 and $\bar{5}$ representations in $10_H = 5_H \oplus \bar{5}_H$ and $10'_H = 5'_H \oplus \bar{5}'_H$ (see e.g. Ref. [45]),

$$\begin{aligned} H_u &\subset \alpha_u \, 5_H + \beta_u \, 5'_H \,, \\ H_d &\subset \alpha_d \, \bar{5}_H + \beta_d \, \bar{5}'_H \,, \end{aligned} \tag{2.5}$$

with arbitrary coefficients $\alpha_{u,d}$ and $\beta_{u,d}$. The CMM model makes the restriction $\beta_u = 0 = \alpha_d$, so that H_u is entirely contained in 10_H and separated from H_d in $10'_H$. In this case, Y_u and Y_ν are generated by the symmetric Y_{10}, whereas Y_d and Y_e stem from the effective coupling Y_{45}, which has arbitrary flavour structure. To see how the Yukawa terms in SO(10) are related to low-energy couplings, we write down the CMM superpotential after SO(10) $\to$ SU(5) breaking,

$$W^Y_{\text{CMM}} = \sqrt{2}\,\tfrac{1}{4} Y^{ij}_{10} \, 10_i 10_j 5_H + \sqrt{2}\, Y^{ij}_{45} \frac{v_{10}}{M_{\text{Pl}}} 10_i \bar{5}_j \bar{5}'_H \\ - \sqrt{2}\, Y^{ij}_{10} \, (\bar{5}_i 1_j + 1_i \bar{5}_j)\, 5_H + Y^{ij}_{16} \frac{v_{16}^2}{2 M_{\text{Pl}}} 1_i 1_j \,. \tag{2.6}$$

---
[1]This choice can be motivated by assigning the Higgs representations 10_H, $10'_H$, and 45_H quantum numbers of a discrete symmetry.

2.1. The Chang-Masiero-Murayama model

Comparing with the SU(5) superpotential in Eq. (1.32) and the GUT Yukawa relations in Eq. (1.34), we identify

$$Y_u = \sqrt{2}\,(Y_{10})_S \qquad Y_\nu = 2\sqrt{2}\,(Y_{10})_S$$
$$Y_d = Y_e^\top = \frac{v_{10}}{M_{\text{Pl}}} Y_{45} \qquad M_N = \frac{v_{16}^2}{M_{\text{Pl}}} Y_{16}\,. \tag{2.7}$$

In this setup, only the Yukawa couplings of up quarks and Dirac neutrinos are generated from a dimension-four term in SO(10). Thereby a hierarchy between large top and smaller bottom and tau Yukawa couplings is naturally given by the suppression factor $v_{10}/M_{\text{Pl}} \sim 10^{-2}$. For "natural" SO(10) Yukawa couplings $Y_{10}, Y_{45} \sim \mathcal{O}(1)$, one has

$$Y_d, Y_e \approx 10^{-2}\, Y_u\,. \tag{2.8}$$

Thus in the CMM model top-bottom-tau Yukawa unification is relaxed to potential bottom-tau unification. Further, $\tan\beta$ has to be small in order to reproduce the measured top and bottom masses at the electroweak scale, following the relation

$$\frac{m_t}{m_b} = \frac{y_t}{y_b}\tan\beta \sim \mathcal{O}(10^2)\,. \tag{2.9}$$

The third term in the CMM superpotential in Eq. (2.3) is well suited to provide masses for right-handed neutrinos. The ratio $v_{16}^2/M_{\text{Pl}} \sim 10^{15}$ GeV is of the right order of magnitude to implement the seesaw mechanism, cf. Sec. 1.1. Below the SU(5) scale, the flavour structure presented so far generally develops a "substructure". Down-quark-lepton Yukawa unification is corrected by additional higher-dimensional contributions from an SU(5)-breaking vev v_5 in 45_H,

$$Y_d = Y_e^\top + 5\frac{v_5}{M_{\text{Pl}}} \widetilde{Y}_{45}\,. \tag{2.10}$$

Due to the suppression with v_5/M_{Pl}, these corrections significantly affect only the first two generations of fermions. They have marginal effects on flavour-changing processes involving the second and third generation, which are in the focus of the CMM model. We therefore neglect Yukawa corrections from the GUT scale in this chapter, but already refer to Chapter 3, where they will be our main subject of interest when studying neutrino mixing in light down-squark FCNC. The structure of symmetry breaking in the Yukawa sector of the CMM model is summarized in Fig. 2.1.

The CMM model makes one crucial assumption concerning flavour mixing: The Yukawa couplings Y_{10} and Y_{16} are simultaneously diagonal. In the U basis, where up quarks are in their mass eigenstates, the CMM superpotential reads

$$W_{\text{CMM}}^U = 16_i \widehat{Y}_{10}^{ii} 16_i 10_H + 16_i (V_q^* \widehat{Y}_{45} V_\ell)^{ij} 16_j \frac{45_H 10'_H}{M_{\text{Pl}}} + 16_i \widehat{Y}_{16}^{ii} 16_i \frac{\overline{16}_H \overline{16}_H}{2M_{\text{Pl}}}\,, \tag{2.11}$$

with $\widehat{Y}^{ii} = \delta_{ij} Y^{ij}$. The alignment of Y_{10} and Y_{16} (and thereby of Y_u, Y_ν, and M_N) can be motivated by a neutrino physics argument: From the measured neutrino oscillation parameters we know that, contrary to up quarks, light neutrinos do not have a strong mass

SO(10)	Y_{10}	Y_{45}	Y_{16}
$\Downarrow \langle 1 \rangle_{45} = v_{10}$	$\Downarrow$	$\Downarrow$	$\Downarrow \langle 1 \rangle_{\overline{16}} = v_{16}$
SU(5)	$Y_5 = \sqrt{2}\, Y_{10},$ $Y_\nu = 2\sqrt{2}\,(Y_{10})_S$	$Y_{\bar 5} = \dfrac{v_{10}}{M_{\rm Pl}} Y_{45}$	$M_N = \dfrac{v_{16}^2}{M_{\rm Pl}} Y_{16}$
$\Downarrow \langle 24 \rangle_{45} = v_5$	$\Downarrow$	$\Downarrow$	$\Downarrow$ seesaw
$G_{\rm SM}$	$Y_u = Y_\nu/2$	$Y_d = Y_e^\top + 5\,\dfrac{v_5}{M_{\rm Pl}} \widetilde Y_{45}$	$M_\nu = v^2 \sin^2\beta\, \dfrac{Y_\nu^2}{M_N}$

Figure 2.1: Symmetry breaking in the CMM model. The Yukawa couplings are defined in Eqs. (2.3), (1.32), and (1.20).

hierarchy. Due to SO(10) unification, however, the neutrino Dirac Yukawa coupling Y_ν exhibits the same strong hierarchy as the up-quark coupling Y_u. Therefore the mass matrix M_N must have a respective double hierarchy in order to compensate for Y_ν^2 in the seesaw mass formula in Eq. (1.19). This pattern is ensured by the ansatz of simultaneously diagonal Yukawa couplings Y_{10} and Y_{16}. Now, flavour mixing can be cleanly studied by exploring the second term in the superpotential. Since this term gives masses to down quarks and charged leptons only, the flavour structure is SU(5)-like, as introduced in Sec. 1.3. Below the GUT scale, the CMM superpotential is given by

$$W^U_{\rm CMM} = Q\,\widehat Y_u\, u^c H_u - Q\,(V_q^*\,\widehat Y_d\, V_\ell)\, d^c H_d - e^c\,(V_q^*\,\widehat Y_e\, V_\ell)\, L H_d - L\,\widehat Y_\nu\, N H_u + \frac{1}{2}\widehat M_N\, N N\,, \qquad (2.12)$$

which is equivalent to the SU(5) superpotential in Eq. (1.36) apart from the alignment of Y_u, Y_ν, and M_N. We identify the matrices V_q and V_ℓ that diagonalize Y_{45},

$$V_q = V_{\rm CKM}, \qquad V_\ell = \Theta_L^\dagger\, V_{\rm PMNS}\, \Theta_R^\dagger\,, \qquad (2.13)$$

and parametrize V_ℓ as in Eq. (1.38). The fact that $V_\ell = V_{\rm PMNS}$ up to phases is due to the alignment of Y_u and M_N. This setup further allows us to identify the rotation matrix of right-handed down (s)quarks, defined in Eq. (1.7), as

$$R_d = V_\ell^\top\,. \qquad (2.14)$$

We benefit from the framework introduced in Sec. 1.3 to describe flavour violation due to large atmospheric neutrino mixing in the down-squark sector. In the super-CKM basis, the mass matrix of right-handed down squarks is given by

$$(M_{\tilde d}^2)^{\rm sCKM} = R_d^\dagger\, (M_{\tilde d}^2)^U\, R_d = m_{\tilde d}^2 \cdot \begin{pmatrix} 1 & 0 & 0 \\ 0 & 1 - \frac{\Delta_{\tilde d}}{2} & -\frac{\Delta_{\tilde d}}{2} e^{-i\phi_{B_s}} \\ 0 & -\frac{\Delta_{\tilde d}}{2} e^{i\phi_{B_s}} & 1 - \frac{\Delta_{\tilde d}}{2} \end{pmatrix}, \qquad (2.15)$$

with the mass splitting $\Delta_{\tilde{d}}$ of $\mathcal{O}(1)$ and the complex phase $\phi_{B_s} = \alpha_4 - \alpha_5$, cf. Eq. (1.45). These sources of flavour and CP violation, quantified by $\Delta_{\tilde{d}}$ and ϕ_{B_s}, are the low-energy imprints of Grand Unification in the CMM model: SU(5) symmetry unifies down (s)quarks and (s)leptons and thus translates large atmospheric neutrino mixing into $\tilde{b}_R - \tilde{s}_R$ transitions. SO(10) symmetry renders this unification effect visible by inducing the large splitting $\Delta_{\tilde{d}}$ in the down-squark mass matrix. We conclude this section by a summary of the main features of the CMM model:

- The CMM model is a phenomenology-orientated perturbative supersymmetric SO(10) model.

- The masses of up quarks are generated by a symmetric Yukawa term of mass dimension four. Down-quark and charged-lepton masses stem from the effective coupling of a dimension-five term, therefore being suppressed by $M_{10}/M_{\text{Pl}} \sim 10^{-2}$. The seesaw mechanism is implemented via a second dimension-five term, leading to a suitable mass scale for right-handed neutrinos of $\mathcal{O}(10^{15}\text{ GeV})$.

- The Yukawa coupling of the dimension-four term is simultaneously diagonal with the mass matrix of right-handed neutrinos, which translates light neutrino mixing directly into mixing among right-handed down (s)quarks.

- Supersymmetry is assumed to be broken at the Planck scale in a flavour-blind way, such that soft masses and trilinear couplings are universal at M_{Pl}. The CMM model is thereby minimally flavour-violating. Without this constraint, flavour mixing effects from Grand Unification could not be distinguished in the arbitrary structure of the soft mass matrices.

2.2 Constraints on the parameter space

One crucial virtue of supersymmetric Grand Unification is its predictivity of masses and couplings, which are free parameters in the MSSM. Namely the vast number of degrees of freedom introduced in the course of SUSY breaking is largely reduced by means of unification relations at the GUT scale. In the CMM model, only six free parameters related to supersymmetry are sufficient to describe the entire spectrum of sparticle masses and couplings at low energy scales. They can be chosen as the gluino mass $M_3 \equiv m_{\tilde{g}}$, cf. Eq. (1.26), the masses of first-generation $\tilde{u}_R$ and $\tilde{d}_R$ squarks $m_{\tilde{u}}$ and $m_{\tilde{d}}$, the ratio of the (1,1) elements of the down trilinear and Yukawa couplings in the super-CKM basis $a_d = (\widehat{A}_d)^{11}/(\widehat{Y}_d)^{11}$, the phase of the μ parameter $\arg(\mu)$, and the ratio of the two Higgs-doublet vevs $\tan\beta$. This set of parameters is used as an input at the electroweak scale and subsequently translated, passing the scales of SU(5) and SO(10) unification, up to the Planck scale via the RGE. By making use of the universality and unification conditions, the parameters are evolved down back to the electroweak scale, spreading into the various SUSY parameters when symmetries are broken at M_{10} and M_{GUT}. This procedure has been implemented into a *Mathematica* program by the authors of Ref. [47].[2] Therein the first- and second-generation Yukawa couplings are set to zero. Higgs masses are assumed to be universal at the Planck scale and equal to the sfermion masses; gaugino masses are set equal above the GUT scale,

$$M_{\text{Pl}}: \quad m_{10_H}^2 = m_{10'_H}^2 = m_{45_H}^2 = m_{16_H}^2 = m_{\overline{16}_H}^2 = m_0^2,$$
$$M_1 = M_2 = M_3 = m_{\tilde{g}}.$$
(2.16)

Thus all soft parameters are fixed from the six input parameters after m_0 and a_0 are determined by evolving $m_{\tilde{u}}$, $m_{\tilde{d}}$, and a_d up to the Planck scale.[3] In the code the light fermion masses are set equal, $m_{\tilde{u}} = m_{\tilde{d}}$, such that the effective parameter space consists of five inputs at M_Z,

$$m_{\tilde{g}}, \quad m_{\tilde{d}}, \quad a_d, \quad \arg(\mu), \quad \tan\beta.$$
(2.17)

This set of CMM input parameters resembles the parameter space of specific SUSY scenarios inspired by minimal supergravity (mSUGRA) like the constrained MSSM (CMSSM). In the CMSSM, the SUSY-breaking parameters are taken to be universal at the GUT scale. The scenario is thus determined by five SUSY inputs: The gaugino and scalar masses $m_{1/2}$ and m_0, the trilinear coupling A, the sign of the μ parameter, and $\tan\beta$ [51]. By lowering the scale of flavour-blind SUSY breaking from M_{Pl} to M_{GUT}, however, one looses the relations between quark and lepton flavours. Consequently, such scenarios do not probe the effects of Grand Unification and lead to a different phenomenology. Still, they are popular in experimental analyses of the MSSM, since they allow to define so-called benchmarks, i.e. fixed characteristic points in parameter space, with a small set of inputs only. It would be a useful attempt to establish equivalent benchmark points based on the CMM model. These would allow to distinguish the phenomenology of GUTs from the MSSM by comparing the different outcome in both models for the same set of input parameters.

We turn to specify various constraints on the CMM input parameters in Eq. (2.17). The requirement of constructing a perturbative model with a realistic MSSM mass spectrum at low energy scales leads to the following restrictions:

Perturbative top Yukawa coupling The ratio of the top Yukawa and gauge couplings exhibits an infrared fixed point in the RGE of SU(5) and SO(10). While the small energy range with SU(5) running doesn't have large impact on the top Yukawa coupling, the SO(10) fixed-point solution acts as a "perturbativity barrier". If the fixed point is reached at M_{10} for a critical value y_t^c, the ratio y_t^c/g stays constant at higher scales. For larger values $y_t(M_{10}) > y_t^c$, the top Yukawa coupling blows up above the SO(10) scale [37]. Via

---
[2] In the code the RGE is executed twice, taking the output parameters of the first run as input for the second run to ensure that the resulting low-scale parameters converge to the "true" values. The program is based on 2-loop running in the MSSM [48], using the $\overline{DR}$ scheme, and on 1-loop running in SU(5) [49] and SO(10) [47].

[3] Both $m_{\tilde{u}}$ and $m_{\tilde{d}}$ are needed to fix the D-term scalar mass splitting [47,50].

2.2. Constraints on the parameter space

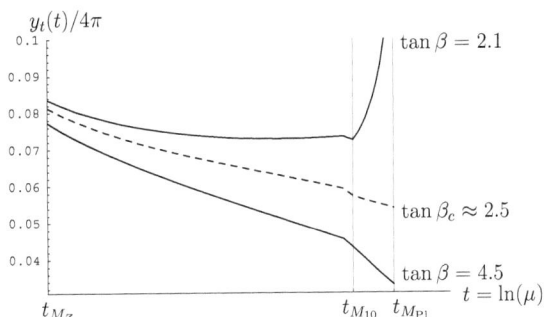

Figure 2.2: Renormalization group evolution of the top Yukawa coupling y_t on the logarithmic energy scale $t = \ln(\mu)$. The dashed line corresponds to the SO(10) fixed-point solution y_t^c (see text).

the RGE, $y_t(M_{10})$ is linked to the corresponding value at the electroweak scale, $y_t(M_Z) = m_t(M_Z)/(v \cdot \sin \beta)$. In order to keep the top Yukawa coupling perturbative at high scales, one thus has to judiciously choose the input value of $\tan \beta$. From Fig. 2.2 [52] one reads off that $y_t < y_t^c$ for $\tan \beta > \tan \beta_c \approx 2.5$, which assures the required perturbativity of y_t up to the Planck scale.

An upper bound on $\tan \beta$ is inherent in the structure of the CMM superpotential. Since the bottom Yukawa coupling is suppressed by $v_{10}/M_{\text{Pl}} \sim 10^{-2}$ with respect to y_t, Eq. (2.9) implies that naturally $\tan \beta \lesssim 10$. The natural range for $\tan \beta$ in the CMM model is thus given by

$$2.5 \lesssim \tan \beta \lesssim 10. \tag{2.18}$$

Flavour effects in the CMM model will be maximal for small $\tan \beta$, which corresponds to a large top Yukawa coupling driving the RGE of soft masses.

Vacuum stability The existence of soft trilinear A terms in the MSSM Lagrangean leads in general to charge- and colour-breaking minima of the scalar potential. In order to preserve the Standard-Model gauge symmetry in the MSSM vacuum state, one needs to impose an upper bound on the trilinear couplings. In particular, the CMM input parameter $a_d = \widehat{A}_d^{11}/\widehat{Y}_d^{11}$ is restricted to fulfill the stability bound [53]

$$|a_d|^2 \leq 3\,(m_{\tilde{Q}}^2 + m_{\tilde{d}}^2 + m_{H_d}^2) \equiv |a_d|_{\max}^2. \tag{2.19}$$

From parameter scans one learns that for $m_{\tilde{d}}/m_{\tilde{q}} < 5$ the normalized stability bound in the CMM model reads $|a_d|_{\max}/m_{\tilde{d}} < 3$. This quantity depends only little on other input parameters. By setting

$$|a_d|/m_{\tilde{d}} < 2.6, \tag{2.20}$$

the stability bound is fulfilled in all regions of the CMM parameter space that are not excluded by other constraints.

Chargino and neutralino masses Due to the universality of gaugino masses above $M_{\rm GUT}$, the gluino mass $m_{\tilde{g}}$ is constrained from experimental lower bounds on neutralino and chargino masses [54]. Especially the chargino mass bound sets a lower limit

$$m_{\tilde{g}} \gtrsim 300\,{\rm GeV} \qquad (2.21)$$

for down-squark masses $m_{\tilde{d}} = \mathcal{O}(1\,{\rm TeV})$. This bound depends only marginally on the other CMM input parameters. Since GUT effects turn out to be most significant for light gluinos, the lower bound on $m_{\tilde{g}}$ limits the magnitude of flavour violation in the CMM model.

Positive sfermion masses Similarly to A terms, negative soft sfermion mass parameters in the scalar potential lead to unwanted charge and colour breaking in the vacuum state. In the CMM model this problem occurs if the large top Yukawa coupling drives the third-generation sfermion masses to negative values at the electroweak scale. The requirement of positive sfermion masses thereby excludes regions in the parameter space where the down-squark mass splitting $\Delta_{\tilde{d}}$ is very large, limiting again the size of flavour-violating effects.

Lightest Higgs mass In the MSSM the mass of the lightest Higgs boson h^0 at tree level is bounded from above by

$$M_{h^0}^{\rm tree} \lesssim M_Z |\cos(2\beta)| \leq 91.2\,{\rm GeV}\,. \qquad (2.22)$$

On the contrary, the experimental lower limit on M_{h^0} for small $\tan\beta$ is close to the Higgs mass bound in the Standard Model [55],

$$M_{h^0}^{\rm exp} \geq 114.4\,{\rm GeV}\,. \qquad (2.23)$$

This contradiction can be resolved by taking into account perturbative corrections to the tree-level Higgs mass. The main contribution stems from top-quark-squark loops [56–58]. For large top squark masses $m_{\tilde{t}}$ corrections are large and positive, so that it is possible to exceed the experimental lower bound. Since large $m_{\tilde{t}}$ implies small squark mass splitting $\Delta_{\tilde{d}}$, the Higgs mass constraint sets a strong upper bound on flavour effects in the CMM model. From Eq. (1.44) we see that small $\Delta_{\tilde{d}}$ requires a moderate top Yukawa coupling $y_t = m_t/(v \cdot \sin\beta)$. Consequently, one needs to balance $\tan\beta$ in order to maximize flavour effects and simultaneously fulfill the constraint on M_{h^0},

small $\tan\beta$: small M_{h^0}, large flavour effects
versus
larger $\tan\beta$: large M_{h^0}, small flavour effects.

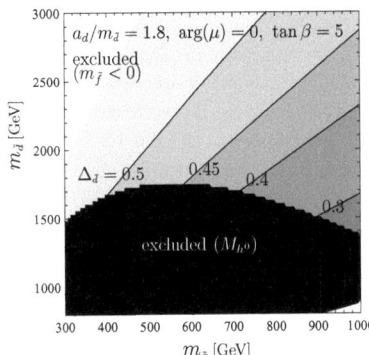

Figure 2.3: Down-squark mass splitting $\Delta_{\tilde{d}}$ as a function of $m_{\tilde{g}}$ and $m_{\tilde{d}}$. White: Negative sfermion masses. Black: Excluded by lower bound on light Higgs mass M_{h^0}.

Apart from small y_t, the Higgs mass corrections are enhanced by a large trilinear coupling A_t, which via unification translates into a large input a_d. On the other hand, for large a_d the mass splitting $\Delta_{\tilde{d}}$ increases, see Eq. (1.44). This leads to small $m_{\tilde{t}}$ and thereby plays against the enhancement of the Higgs mass from suppressed y_t. To reach significant flavour effects, one thus has to moderately increase a_d, such that $\Delta_{\tilde{d}}$ is preferably large without spoiling the corrections to the light Higgs mass.

The constraints on the CMM parameter space from the light Higgs and sfermion mass bounds are illustrated in Fig. 2.3. The plot shows the down-squark mass splitting $\Delta_{\tilde{d}}$ in the $m_{\tilde{g}} - m_{\tilde{d}}$ plane for the specific set of input parameters $a_d/m_{\tilde{d}} = 1.8$, $\arg(\mu) = 0$, and $\tan\beta = 5$. For this choice of inputs the other abovementioned constraints are fulfilled, while still preserving a wide region in the CMM parameter space where flavour effects can be large. Notice that $\tan\beta$ is increased over $\tan\beta_c \approx 2.5$ in order to fulfill the Higgs mass bound for soft masses $m_{\tilde{g}}$ and $m_{\tilde{d}}$ below 2 TeV. Still, the down-squark mass splitting can reach values up to $\Delta_{\tilde{d}} = 0.55$ because of the sizeable value of a_d. This scenario will serve as our framework in the subsequent phenomenological analysis of CMM flavour effects.

2.3 Phenomenology of down-quark-lepton unification

At the end of Sec. 1.3 we already mentioned phenomenological implications of large atmospheric neutrino mixing in $b - s$ transitions. Within the CMM model, the following observables have hitherto been studied quantitatively: In Ref. [37] the impact of $\tilde{b}_R - \tilde{s}_R$ mixing on the mass difference ΔM_s and CP phase ϕ_s in $B_s - \overline{B}_s$ mixing has been examined. A comprehensive study of $B \to X_s \gamma$, ΔM_s, and $\tau \to \mu\gamma$ is given by Ref. [47], resulting in constraints on the CMM parameter space. Here we focus on neutrino mixing effects in meson mixing. In Sec. 2.5, we reinvestigate $B_s - \overline{B}_s$ mixing. Both ΔM_s and ϕ_s having recently been measured at the B factories, we confront the CMM contribution to these observables with experiment. In Chapter 3, we extend the CMM model to study effects of large atmospheric neutrino mixing in $K - \overline{K}$ and $B_d - \overline{B}_d$ mixing.

The characteristic effect of down-quark-lepton unification in the CMM model becomes manifest in FCNC involving down squarks. In an MSSM with generic flavour structure in soft SUSY-breaking terms, neutral flavour-changing vertices can be generated by quark-squark-gluino and quark-squark-neutralino couplings. We will concentrate on the gluino vertex, which is proportional to the strong coupling g_3 and therefore numerically dominant over the weak neutralino couplings. The down-quark-squark-gluino vertex in the CMM model is displayed in Fig. 2.4. Contrarily to the MSSM with generic soft parameters, where flavour mixing in this coupling is arbitrary, in the CMM model flavour mixing among down (s)quarks is determined by $R_d \sim V_\text{PMNS}$ and thereby directly linked to flavour mixing in the lepton sector. Based on these considerations, we proceed and introduce the formalism to make this "CMM effect" visible in meson mixing.

Meson mixing

A neutral meson P is distinguished from its antimeson $\overline{P}$ by a quantum number F characterizing its quark flavour content. In the case of neutral K and B_d/B_s mesons this is strangeness S and beauty B with, for instance, $S(K = d\bar{s}) = -1$ and $S(\overline{K} = \bar{d}s) = +1$. Oscillations between P and $\overline{P}$ are thus induced by interactions that change the flavour quantum number by two units, $\Delta F = 2$. In the Standard Model, such flavour violation is possible through weak interactions, namely the box diagram displayed in Fig. 2.5(a) for the kaon system. In the CMM model, competitive additional contributions from large neutrino mixing arise via a squark-gluino box, see Fig. 2.5(b). The corresponding diagrams for $B_d - \overline{B}_d$ and $B_s - \overline{B}_s$ mixing are obtained by adapting the external quark flavours.

The time evolution of a $P - \overline{P}$ system in the Wigner-Weisskopf approximation [59, 60] is

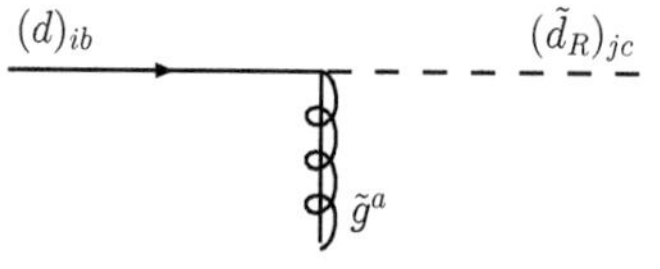

$$\begin{array}{c}(d)_{ib} \qquad\qquad (\tilde{d}_R)_{jc}\\ \rule{4cm}{0.4pt}\\ \tilde{g}^a\end{array} \quad = i\sqrt{2}\, g_3 T^a_{cb}(R_d^*)_{ji}\, P_R$$

Figure 2.4: Flavour-violating down-quark-squark-gluino vertex in the CMM model. The generators T^a in the fundamental representation of $SU(3)_C$ determine the colour structure. $P_R = (1 + \gamma_5)/2$ projects on the right-handed component of the down-quark field d. The matrix $R_d = V_\ell^\top$ is given in Eq. (1.38).

2.3. Phenomenology of down-quark-lepton unification

Figure 2.5: Dominant short-distance contributions to M_{12}^K (a) in the SM; (b) in the CMM extension. k and m are flavour indices. Diagrams with flipped $W \leftrightarrow q$ and $\tilde{g} \leftrightarrow \tilde{d}$ are not displayed.

conveniently described by

$$i\frac{d}{dt}\psi(t) = H\psi(t), \qquad \psi(t) = \begin{pmatrix} |P(t)\rangle \\ |\overline{P}(t)\rangle \end{pmatrix},$$

$$\text{with} \quad H = M - \frac{i}{2}\Gamma = \begin{pmatrix} M_{11} - \frac{i}{2}\Gamma_{11} & M_{12} - \frac{i}{2}\Gamma_{12} \\ M_{12}^* - \frac{i}{2}\Gamma_{12}^* & M_{22} - \frac{i}{2}\Gamma_{22} \end{pmatrix}. \quad (2.24)$$

M and Γ are hermitean matrices describing $P - \overline{P}$ transitions via virtual and physical intermediate states, respectively. Due to CPT symmetry, one has $M_{11} = M_{22}$ and $\Gamma_{11} = \Gamma_{22}$. The physical eigenstates of this system are obtained by diagonalization to be

$$|P_1(t)\rangle = p|P(t)\rangle + q|\overline{P}(t)\rangle, \quad |P_2(t)\rangle = p|P(t)\rangle - q|\overline{P}(t)\rangle, \quad \frac{q}{p} = \sqrt{\frac{M_{12}^* - \frac{i}{2}\Gamma_{12}^*}{M_{12} - \frac{i}{2}\Gamma_{12}}}, \quad (2.25)$$

with the corresponding eigenvalues

$$M_1 - \frac{i}{2}\Gamma_1 = M_{11} - \frac{i}{2}\Gamma_{11} + \frac{q}{p}(M_{12} - \frac{i}{2}\Gamma_{12}),$$

$$M_2 - \frac{i}{2}\Gamma_2 = M_{11} - \frac{i}{2}\Gamma_{11} - \frac{q}{p}(M_{12} - \frac{i}{2}\Gamma_{12}), \quad (2.26)$$

We use the phase convention $CP|P(t)\rangle = -|\overline{P}(t)\rangle$ and define the relative phase between M_{12} and Γ_{12},

$$\phi \equiv \arg\left(\frac{M_{12}}{-\Gamma_{12}}\right). \quad (2.27)$$

Going back to Eq. (2.25), we see that the physical states are CP eigenstates only if $|q/p| = 1$, or equivalently $\phi = 0$. In this case, $|P_1(t)\rangle$ and $|P_2(t)\rangle$ are CP-odd and CP-even, respectively,

$$CP|P_{1,2}(t)\rangle = \eta_{1,2}|P_{1,2}(t)\rangle, \qquad \eta_{1,2} = -1, +1. \quad (2.28)$$

The mass and width differences between the two eigenstates are defined by

$$\Delta M \equiv M_1 - M_2 = 2\operatorname{Re}\!\left(\frac{q}{p}(M_{12} - \tfrac{i}{2}\Gamma_{12})\right),$$
$$\Delta\Gamma \equiv \Gamma_1 - \Gamma_2 = -4\operatorname{Im}\!\left(\frac{q}{p}(M_{12} - \tfrac{i}{2}\Gamma_{12})\right). \tag{2.29}$$

For the K and B systems, approximations to these expressions can be derived by means of empirical observations. In both $B_d - \overline{B}_d$ and $B_s - \overline{B}_s$ mixing one has $|\Gamma_{12}| \ll |M_{12}|$, leading to the mass differences

$$\Delta M_d = 2\,|M_{12}^d|, \qquad \Delta M_s = 2\,|M_{12}^s|. \tag{2.30}$$

The situation is different in $K - \overline{K}$ mixing, where $\Delta\Gamma$ and ΔM are of the same order of magnitude. Still, the phase difference ϕ between $|\Gamma_{12}|$ and $|M_{12}|$ is very small, so that the short-distance contributions to the mass difference from Fig. 2.5 are again well approximated by

$$\Delta M_K = 2\,|M_{12}^K|. \tag{2.31}$$

However, ΔM_K receives sizeable long-distance contributions from the exchange of light mesons, which are hard to estimate. Therefore, despite its precise measurement, ΔM_K plays a minor role in the search for new-physics contributions. Comprehensive discussions of meson mixing can be found in Refs. [61, 62].

The mass differences probe CMM effects in the mixing element $|M_{12}|$. To study the CP-violating properties of CMM contributions in meson mixing, one has to focus on the mixing phase $\phi_M = \arg(M_{12})$. In the B_d and B_s systems, the time-dependent CP asymmetries A_{CP} in the decays $B_d \to J/\psi K_S$ and $B_s \to J/\psi\phi$ are well-suited observables for this purpose. A_{CP} is a measure of CP violation in the interference between meson mixing and decay amplitudes. Since neither of both decays exhibits direct CP violation, the asymmetry provides clean access to the mixing phase ϕ_M. For negligible $\Delta\Gamma$ and Γ_{12} the time-dependent CP asymmetry is given by

$$\begin{aligned}
A_{CP}(f) &\equiv \frac{\Gamma(\overline{B}(t) \to f) - \Gamma(B(t) \to f)}{\Gamma(\overline{B}(t) \to f) + \Gamma(B(t) \to f)} \\
&= \operatorname{Im}\!\left(\frac{q}{p}\cdot\frac{\overline{A}(f)}{A(f)}\right)\sin(\Delta M t) = -\eta_f \sin(\phi_M)\sin(\Delta M t),
\end{aligned} \tag{2.32}$$

where $\overline{A}(f)/A(f)$ is the ratio of the decay amplitudes for $\overline{B} \to f$ and $B \to f$. The sign of A_{CP} depends on the CP quantum number η_f of the respective final state. Let us first consider the CP asymmetry in $B_d \to J/\psi K_S$, where the final state $f = J/\psi K_S$ is CP-odd, i.e. $\eta_{J/\psi K_S} = -1$. The decay process is illustrated in Fig. 2.6. In the Standard Model, ϕ_M

2.3. Phenomenology of down-quark-lepton unification

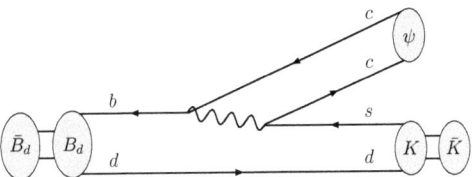

Figure 2.6: Decay process $\bar{B}_d \to J/\psi K_S$. The $\bar{B}_d$ meson can decay into the CP eigenstate $J/\psi K_S$ either via mixing or directly.

is closely related to the angle β of the unitarity triangle, following

$$\frac{q}{p} \cdot \frac{\bar{A}(J/\psi K_S)}{A(J/\psi K_S)} = \sqrt{\frac{(M_{12}^d)^*}{M_{12}^d}} \cdot \frac{\bar{A}(J/\psi K_S)}{A(J/\psi K_S)} = e^{-i\phi_M} \cdot \frac{\bar{A}(J/\psi K_S)}{A(J/\psi K_S)}$$
$$= -\eta_{J/\psi K_S} \cdot \frac{V_{td}V_{tb}^*}{V_{td}^*V_{tb}} \cdot \frac{V_{cb}V_{cs}^*}{V_{cb}^*V_{cs}} \cdot \frac{V_{cs}V_{cd}^*}{V_{cs}^*V_{cd}} = -e^{-2i\beta}, \quad \beta \equiv \arg\left(-\frac{V_{td}^*V_{tb}}{V_{cd}^*V_{cb}}\right). \tag{2.33}$$

The three factors of CKM elements stem from $B_d - \bar{B}_d$ mixing, the $B_d \to J/\psi K_S$ decay, and subsequent $K - \bar{K}$ mixing. The minus sign is due to our phase convention $CP|P\rangle = -|\bar{P}\rangle$. In the standard CKM phase convention one identifies $\phi_M = 2\beta$, leading to

$$A_{CP}^{\text{SM}}(J/\psi K_S) = \sin(2\beta)\sin(\Delta M_d t). \tag{2.34}$$

Consequently, in the Standard Model the CP asymmetry in $B_d \to J/\psi K_S$ measures the UT angle β. New-physics contributions to $B_d - \bar{B}_d$ mixing will show up in A_{CP} as a phase shift ϕ_d^Δ. Since new physics in the decay amplitude is in general negligible compared to the SM contribution, the CP asymmetry reads

$$S_{J/\psi K_S} \equiv \sin(2\beta + \phi_d^\Delta) = \text{Im}\left(\frac{M_{12}^d}{|M_{12}^d|}\right), \quad \phi_d^\Delta \equiv \arg\frac{M_{12}^d}{(M_{12}^d)^{\text{SM}}}. \tag{2.35}$$

The decay $B_s \to J/\psi\phi = (c\bar{c})(s\bar{s})$ differs from $B_d \to J/\psi K_S$ by merely changing the spectator quark in the B meson from d to s, cf. Fig. 2.6 again. Analogously, one finds the relevant quantity for the CP asymmetry expressed in terms of CKM elements,

$$\frac{q}{p} \cdot \frac{\bar{A}(J/\psi\phi)}{A(J/\psi\phi)} = -\eta_{J/\psi\phi} \cdot \frac{V_{ts}V_{tb}^*}{V_{ts}^*V_{tb}} \cdot \frac{V_{cb}V_{cs}^*}{V_{cb}^*V_{cs}} = \eta_{J/\psi\phi} \cdot e^{2i\beta_s}, \quad \beta_s \equiv -\arg\left(-\frac{V_{ts}^*V_{tb}}{V_{cs}^*V_{cb}}\right), \tag{2.36}$$

and consequently $\phi_M = -2\beta_s$. Even though the final state $f = J/\psi\phi$ is not a CP eigenstate, it is possible to extract the CP asymmetry in $B_s \to J/\psi\phi$ from the angular distribution of the decay products. We thus can compare new-physics contributions to $A_{CP}(J/\psi\phi)$

$\Delta M_K^{\text{exp}} = (3.483 \pm 0.006) \cdot 10^{-12}$ MeV [54]	$\|\epsilon_K\|^{\text{exp}} = (2.229 \pm 0.012) \cdot 10^{-3}$ [54]
$\Delta M_d^{\text{exp}} = (3.337 \pm 0.033) \cdot 10^{-10}$ MeV [54]	$S_{J/\psi K_S}^{\text{exp}} = 0.671 \pm 0.024$ [63]
$\Delta M_s^{\text{exp}} = (117.0 \pm 0.8) \cdot 10^{-10}$ MeV [54]	$\phi_s^{\text{exp}} = (-0.77^{+0.29}_{-0.37}) \cup (-2.36^{+0.37}_{-0.29})$ rad [63]

Table 2.1: Experimental values of meson mixing observables.

and $A_{CP}(J/\psi K_S)$ via

$$S_{J/\psi\phi} \equiv \sin(-2\beta_s + \phi_s^\Delta) = \text{Im}\left(\frac{M_{12}^s}{|M_{12}^s|}\right), \qquad \phi_s^\Delta \equiv \arg\frac{M_{12}^s}{(M_{12}^s)^{\text{SM}}}. \qquad (2.37)$$

In the Standard Model, $S_{J/\psi\phi}$ is tiny due to the small mixing phase $\phi_M = -2\beta_s \approx -0.04$. Since Γ_{12} doesn't exhibit a significant CP phase, one generally identifies

$$\phi_M = \phi_s = \arg\left(\frac{M_{12}^s}{-\Gamma_{12}^s}\right). \qquad (2.38)$$

The measurements of the observables discussed in this paragraph are given in Tab. 2.1.

The meson mixing element M_{12} is generally calculated in the framework of a low-energy effective theory that separates weak interactions of $\mathcal{O}(M_W)$ between the mesons from strong interactions of $\mathcal{O}(1\,\text{GeV})$. The corresponding effective weak Hamiltonian

$$\mathcal{H}_{\text{eff}}^{\Delta F=2} = \frac{G_F^2 M_W^2}{16\pi^2} \sum_i C_P^i(\mu_P) Q_P^i(\mu_P), \qquad G_F = \frac{\sqrt{2}\,g_2^2}{8M_W^2}, \qquad (2.39)$$

describes weak interactions at low energy scales $\mu_B \sim M_B$, $\mu_K \lesssim m_c$ in terms of a series of local operators $Q^i(\mu_P)$, weighted with Wilson coefficients $C^i(\mu_P)$. The mixing M_{12}^P is obtained by calculating the matrix element of $\mathcal{H}_{\text{eff}}^{\Delta F=2}$ between meson and antimeson states,

$$2M_P \cdot M_{12}^P = \langle P | \mathcal{H}_{\text{eff}}^{\Delta F=2} | \bar{P} \rangle = \frac{G_F^2 M_W^2}{16\pi^2} \sum_i C_P^i(\mu_P) \langle P | Q_P^i(\mu_P) | \bar{P} \rangle, \qquad (2.40)$$

with $M_P = (M_{P_1} + M_{P_2})/2$. The Wilson coefficients $C^i(\mu_P)$ are derived from matching the calculated box diagrams in Fig. 2.5 to the effective theory at $\mu = \mathcal{O}(M_W)$ and then evolving down to $\mu = \mu_P$ by means of the renormalization group. This formalism stores perturbative effects of $\mathcal{O}(\mu > \mu_P)$ in the Wilson coefficients to separate them from non-perturbative contributions of $\mathcal{O}(\mu < \mu_P)$ in the matrix elements $\langle P | Q^i(\mu_P) | \bar{P} \rangle$. The latter can be obtained from non-perturbative methods like lattice calculations. An extensive introduction into the theory of effective weak interactions is given in Ref. [64].

Standard-model contributions In the Standard Model, the contributions to meson mixing from the weak box diagrams in Fig. 2.5(a) are described by only one effective operator for each of the meson systems,

$$Q_K = (\bar{d}_L \gamma_\mu s_L)(\bar{d}_L \gamma^\mu s_L), \qquad Q_{B_q} = (\bar{q}_L \gamma_\mu b_L)(\bar{q}_L \gamma^\mu b_L), \qquad q = d, s. \qquad (2.41)$$

2.3. Phenomenology of down-quark-lepton unification

The corresponding Wilson coefficients are given by

$$C_K(\mu_K) = 4 U_K(\mu_K)\Big[(V_{cd}^*V_{cs})^2 \eta_1 S_0(x_c) \\
+ 2(V_{cd}^*V_{cs})(V_{td}^*V_{ts})\eta_3 S_0(x_c, x_t) + (V_{td}^*V_{ts})^2 \eta_2 S_0(x_t)\Big], \quad (2.42)$$

$$C_{B_q}(\mu_{B_q}) = 4 U_B(\mu_{B_q})(V_{tq}^*V_{tb})^2 \eta_B S_0(x_t).$$

For massless internal quarks the contributions from up, charm, and top loops cancel due to the GIM mechanism, so that there is no meson mixing. In the B_d and B_s systems, only the contribution with heavy top quarks is relevant, while light up and charm quarks are negligible. In the K system, in addition one has to take account of charm and charm-top contributions, because the top loops are suppressed by small CKM elements, cf. Eq. (2.42). The loop functions $S_0(x_q)$ with $x_q = m_q^2/M_W^2$ are given in Appendix A.3. The factors $\eta_{1,2,3}$ and η_B comprise perturbative QCD corrections to these diagrams up to next-to-leading order (NLO) at high scales of $\mathcal{O}(M_W)$ [65,66]. The RGE of the Wilson coefficients at NLO is encoded in the functions

$$U_K(\mu) = \big[\alpha_s^{(3)}(\mu)\big]^{-\frac{2}{9}}\Big[1 + \frac{\alpha_s^{(3)}(\mu)}{4\pi} J_3\Big] \quad \text{and} \quad U_B(\mu) = \big[\alpha_s^{(5)}(\mu)\big]^{-\frac{6}{23}}\Big[1 + \frac{\alpha_s^{(5)}(\mu)}{4\pi} J_5\Big], \quad (2.43)$$

where the high-energy dependence has been absorbed into η_i. $\alpha_s^{(n)}(\mu)$ with $\alpha_s \equiv g_3^2/4\pi$ is the strong coupling in a framework with n active quark flavours. Explicit expressions for J_3 and J_5 are given in Ref. [64]. To compute M_{12}^P from Eq. (2.40), one needs the matrix elements of the operators Q_K and Q_{B_q}. They are parametrized in terms of "bag factors" B_P at low scales $\mu = \mathcal{O}(\mu_P)$,

$$\langle P|Q_P(\mu)|\overline{P}\rangle = \frac{2}{3} M_P^2 F_P^2 B_P(\mu), \quad (2.44)$$

where F_P is the decay constant of the meson P. The scale dependence of $U_P(\mu)$ and $B_P(\mu)$ cancels in the product $\widehat{B}_P = B_P(\mu) U_P(\mu)$. From Eqs. (2.40), (2.42), and (2.44) one composes the mixing elements

$$(M_{12}^K)^{\text{SM}} = \frac{G_F^2 M_W^2}{12\pi^2} M_K F_K^2 \widehat{B}_K \big[(\lambda_{ds}^c)^2 \eta_1 S_0(x_c) + 2(\lambda_{ds}^c)(\lambda_{ds}^t)\eta_3 S_0(x_c, x_t) + (\lambda_{ds}^t)^2 \eta_2 S_0(x_t)\big],$$

$$(M_{12}^q)^{\text{SM}} = \frac{G_F^2 M_W^2}{12\pi^2} M_{B_q} F_{B_q}^2 \widehat{B}_{B_q} (\lambda_{qb}^t)^2 \eta_B S_0(x_t), \qquad \lambda_{ij}^k \equiv V_{ki}^* V_{kj}. \quad (2.45)$$

CMM contributions In the context of the CMM model, the main additional contributions to meson mixing stem from the gluino diagrams in Fig. 2.5(b) with internal $\tilde{d}_R$, $\tilde{s}_R$, and $\tilde{b}_R$ squarks. They introduce the parity-flipped operators

$$Q_K^{\text{CMM}} = \big(\overline{d}_R \gamma_\mu s_R\big)\big(\overline{d}_R \gamma^\mu s_R\big), \qquad Q_{B_q}^{\text{CMM}} = \big(\overline{q}_R \gamma_\mu b_R\big)\big(\overline{q}_R \gamma^\mu b_R\big). \quad (2.46)$$

The Wilson coefficients at the scale of SUSY particles $\mu_S = \mathcal{O}(m_{\tilde{d}}, m_{\tilde{g}})$ are given by

$$C_P^{\text{CMM}}(\mu_S) = \frac{16\pi^2}{G_F^2 M_W^2} \frac{\alpha_s^2(\mu_S)}{2m_{\tilde{g}}^2} \sum_{k,m=1}^{3} (R_d)_{mj}(R_d)^*_{mi}(R_d)_{kj}(R_d)^*_{ki} L_0(x_m, x_k), \qquad (2.47)$$

with flavour indices $(i,j) = (1,2)$ for the K system and $(i,j) = (1,3)$ and $(2,3)$ for the B_d and B_s systems. The loop function $L_0(x_m, x_k)$ with $x_n = m_{\tilde{d}_n}^2/m_{\tilde{g}}^2$ is given in Appendix A.3. The large $(2,3)$ element of the rotation matrix R_d, defined in Eqs. (2.14) and (1.38), leads to sizeable Wilson coefficients $C_{B_s}^{\text{CMM}}$ that compete with the SM ones for B_s mixing. In the K and B_d systems, contributions to the CMM Wilson coefficients arise only from corrections of Yukawa unification and will be studied in Chapter 3. Making use of the mass degeneracy of the first two sfermion generations (cf. Eq. (1.42)) and the unitarity of R_d, Eq. (2.47) simplifies to

$$C_P^{\text{CMM}}(\mu_S) = \frac{16\pi^2}{G_F^2 M_W^2} \frac{\alpha_s^2(\mu_S)}{2m_{\tilde{g}}^2} [(R_d)_{3j}(R_d)^*_{3i}]^2 S^{(\tilde{g})}(x_1, x_3),$$
$$x_1 = m_{\tilde{d}}^2/m_{\tilde{g}}^2, \qquad x_3 = m_{\tilde{d}}^2(1 - \Delta_{\tilde{d}})/m_{\tilde{g}}^2, \qquad (2.48)$$

with the effective loop function

$$S^{(\tilde{g})}(x_1, x_3) = L_0(x_1, x_1) - 2L_0(x_1, x_3) + L_0(x_3, x_3). \qquad (2.49)$$

Since the inputs of the CMM model in Eq. (2.17) are defined at the electroweak scale $\mu_Z = \mathcal{O}(M_Z, M_W, m_t)$, we set $\mu_S = \mu_Z$ and thereby neglect effects due to the RGE of the CMM Wilson coefficients from μ_S down to μ_Z. The evolution to the meson scale μ_P proceeds as in the Standard Model, leading to

$$C_K^{\text{CMM}}(\mu_K) = U_K(\mu_K) \frac{\eta_2}{r} C_K^{\text{CMM}}(\mu_S = \mu_Z), \quad C_B^{\text{CMM}}(\mu_B) = U_B(\mu_B) \frac{\eta_B}{r} C_B^{\text{CMM}}(\mu_Z). \quad (2.50)$$

The factor $r = 0.985$ [66] removes the NLO QCD corrections to the SM loop function $S_0(x_t)$ from η_2 and η_B. The bag parameters for the matrix elements of the CMM operators being the same as in the Standard Model, the CMM contributions to $B_s - \overline{B}_s$ mixing read

$$(M_{12}^s)^{\text{CMM}} = \frac{\alpha_s^2(M_Z)}{6m_{\tilde{g}}^2} \left[\frac{1}{2} e^{-i\phi_{B_s}}\right]^2 M_{B_s} F_{B_s}^2 \widehat{B}_{B_s} \frac{\eta_B}{r} S^{(\tilde{g})}(x_1, x_3). \qquad (2.51)$$

Here the elements of $(R_d)_{3n}$ in Eq. (2.48) have been made explicit assuming tri-bi-maximal neutrino mixing and using the phase ϕ_{B_s} defined below Eq. (2.15). One immediately reads off that CMM contributions are large for small gluino masses $m_{\tilde{g}}$.

Additional supersymmetric contributions Finally, we comment on supersymmetric contributions that are not affected by large neutrino mixing in the CMM model, namely charged-Higgs(H)-quark and chargino(χ)-squark box diagrams. They do not introduce

2.4. Numerical setup

$\kappa_\epsilon = 0.92 \pm 0.02$ [67]	$\|V_{us}\| = 0.2246 \pm 0.0012$	[68]
$F_K = (156.1 \pm 0.8)$ MeV [68]	$\|V_{cb}\| = (41.6 \pm 0.6) \cdot 10^{-3}$	[54]
$\widehat{B}_K = 0.75 \pm 0.07$ [69]	$\|V_{ub}\| = (3.95 \pm 0.35) \cdot 10^{-3}$	[54]
$F_{B_s}\widehat{B}_{B_s}^{1/2} = (270 \pm 30)$ MeV [69]	$\gamma = (70.7^{+5.7}_{-7.0})°$	[s.text]
$\xi \equiv \frac{F_{B_s}\widehat{B}_{B_s}^{1/2}}{F_{B_d}\widehat{B}_{B_d}^{1/2}} = (1.21 \pm 0.04)$ [69]	$\eta_1 = (1.32 \pm 0.32) \left[\frac{1.30\,\text{GeV}}{\overline{m}_c(m_c)}\right]^{1.1}$	[65, 70]
$\overline{m}_c(m_c) = (1.266 \pm 0.014)$ GeV [71]	$\eta_2 = 0.57 \pm 0.01$	[66, 70]
$\overline{m}_t(m_t) = (162.1 \pm 1.2)$ GeV [72, 73]	$\eta_3 = 0.47 \pm 0.05$	[65, 70]
$\alpha_s(M_Z) = 0.1176 \pm 0.0020$ [54]	$\eta_B = 0.551 \pm 0.007$	[66, 74]

Table 2.2: Input parameters for meson mixing observables.

$\sin^2\theta_W = 0.23122$	[54]	$\widehat{m}_t = 171.4$ GeV	[72]
$M_Z = 91.1876$ GeV	[54]	$\widehat{m}_b = 4.914$ GeV	[54]
$\alpha_e(M_Z) = 1/128.93$	[75]	$m_\tau = 1.77699$ GeV	[54]
$\alpha_s(M_Z) = 0.1176$	[54]		

Table 2.3: Input parameters for the RGE of Yukawa couplings and SUSY parameters. $\widehat{m}_t$ and $\widehat{m}_b$ are two-loop pole masses. The precise value of $\widehat{m}_b$ is inessential for the analysis, while $\widehat{m}_t$ matters. $\alpha_e \equiv g_1^2 \cos^2\theta_W/4\pi$ is the electromagnetic coupling.

new operators, and the flavour structure of the corresponding matrix elements is the same as in the Standard Model,

$$(M_{12}^K)^{H+\chi} = \frac{G_F^2 M_W^2}{12\pi^2} M_K F_K^2 \widehat{B}_K \left\{ 2(\lambda_{ds}^c)(\lambda_{ds}^t)\eta_3^H S_H(c,t) + (\lambda_{ds}^t)^2 \eta_2 \left[S_H(t,t) + S_\chi(t,t)\right] \right\},$$
$$(M_{12}^q)^{H+\chi} = \frac{G_F^2 M_W^2}{12\pi^2} M_{B_q} F_{B_q}^2 \widehat{B}_{B_q} (\lambda_{qb}^t)^2 \eta_B \left[S_H(t,t) + S_\chi(t,t)\right]. \quad (2.52)$$

The loop functions $S_H(c,t)$, $S_H(t,t)$, and $S_\chi(t,t)$ are given explicitly in Ref. [22]. The factor $\eta_3^H = 0.21$ [22] denotes leading-order QCD corrections to the charged-Higgs box with virtual flavours (c,t). Numerically, charged-Higgs and chargino contributions are small compared to gluino contributions in the CMM parameter space. We checked explicitly that they can be neglected in our analysis.

2.4 Numerical setup

In order to perform a numerical study of CMM effects in meson mixing, we have to consider three classes of inputs: One, experimental and theoretical quantities needed for the calculation of the various observables. Two, inputs for the RGE of Yukawa couplings and SUSY parameters. Three, parameter sets for the CMM inputs in Eq. (2.17). The CMM phase ϕ_{B_s} remains a free parameter in our analysis. All inputs of class one are reported in Tab. 2.2.

	$m_{\tilde{g}}$ [GeV]	$m_{\tilde{d}}$ [GeV]	$\Delta_{\tilde{d}}$
Set 1	400	2000	0.52
Set 2	700	2000	0.44
Set 3	700	3000	0.51

Table 2.4: CMM parameter sets for fixed $a_d/m_{\tilde{d}} = 1.8$, $\arg(\mu) = 0$, and $\tan\beta = 5$.

The meson decay constants F_P and bag factors $\hat{B}_P$ are taken from lattice calculations. The quantities $\eta_{1,2,3}$ and η_B comprise the NLO QCD corrections to K and B meson mixing. $\overline{m}_c(m_c)$ and $\overline{m}_t(m_t)$ denote the charm and top quark masses in the $\overline{MS}$ renormalization scheme entering the loop functions in SM meson mixing. Inputs related to CKM elements have to be protected from new-physics impact. To this end, we determine the CKM matrix from the elements $|V_{ub}|$, $|V_{cb}|$, $|V_{us}|$, and δ, the CP phase in the standard parametrization, which equals the angle γ of the unitarity triangle to very good accuracy. The three CKM elements are extracted from tree-level decays. We use $|V_{us}| = 0.2246 \pm 0.0012$ [68], the inclusive determination $|V_{cb}| = (41.6 \pm 0.6) \cdot 10^{-3}$ [54], and the average of inclusive and exclusive determinations $|V_{ub}| = (3.95 \pm 0.35) \cdot 10^{-3}$ [54]. The angle γ is determined via $\gamma = \pi - \alpha - \beta = \pi - \alpha^{\text{eff}} - \beta^{\text{eff}}$, with $\beta^{\text{eff}} = \beta + \phi_d^\Delta/2 = (21.1 \pm 0.9)°$ from $S_{J/\psi K_S}$ [63] and $\alpha^{\text{eff}} = \alpha - \phi_d^\Delta/2 = (88.2^{+6.1}_{-4.8})°$ from $B \to \pi\pi$, $\pi\rho$, $\rho\rho$ decays [76]. The dependence on the new-physics phase ϕ_d^Δ cancels out in the sum $\alpha^{\text{eff}} + \beta^{\text{eff}}$, such that $\gamma = (70.7^{+5.7}_{-7.0})°$ is indeed free from new-physics contamination.

Further inputs needed to perform the RGE of the Yukawa couplings and SUSY parameters within the CMM model are given in Tab. 2.3. Besides some quantities related to electroweak symmetry breaking one needs the masses of third-generation quarks. The dominant effects in the RGE stem from the large top Yukawa coupling. Since within the CMM model up-type quarks are not unified with down-type quarks and leptons, the smaller bottom and tau Yukawa couplings are additional inputs to determine the trilinear $A_{d,e}$ couplings, cf. Eq. (1.28). RGE effects from light quarks are neglected. The scales of unification are fixed to

$$M_{\text{GUT}} \simeq 10^{16}\,\text{GeV} \simeq v_5,$$
$$M_{10} = 10^{17}\,\text{GeV} = v_{10}, \quad (2.53)$$
$$M_{\text{Pl}} = 10^{19}\,\text{GeV},$$

where M_{GUT} is defined by the unification of the gauge couplings $g_1' = \sqrt{5/3}\, g_1$ and g_2. Concerning the SUSY input parameters, we preselect a setup in which CMM effects are sizeable, but still allowed within the constraints discussed in Sec. 2.2. Correspondingly, we fix three of the five input parameters in Eq. (2.17),

$$a_d/m_{\tilde{d}} = 1.8, \quad \arg(\mu) = 0, \quad \tan\beta = 5, \quad (2.54)$$

leaving the gluino and down squark masses $m_{\tilde{g}}$ and $m_{\tilde{d}}$ as free parameters. The CMM contributions to meson mixing observables are parametrized by just these soft masses and

2.5. The CP phase in $B_s - \overline{B}_s$ meson mixing

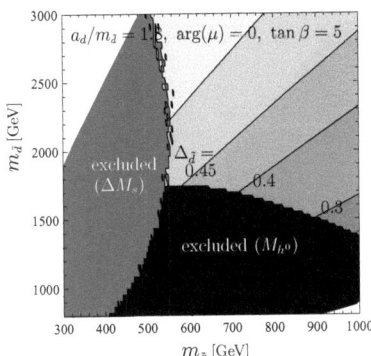

Figure 2.7: 3-sigma constraints on CMM parameter space from ΔM_s for $\phi_{B_s} = 0$ (dark gray). White: Negative sfermion masses. Black: Excluded by lower bound on light Higgs mass M_{h^0}.

the down-squark mass splitting $\Delta_{\tilde{d}}$, cf. Eq. (2.48). In Tab. 2.4, we therefore define three sets distinguished by the inputs for $m_{\tilde{g}}$ and $m_{\tilde{d}}$ to estimate the respective size of CMM effects for specific points in parameter space. The output for $\Delta_{\tilde{d}}$ within each of these sets is given in the last column. All three sets respect the constraints from $B \to X_s\gamma$ and $\tau \to \mu\gamma$ studied in Ref. [47]. Set 1 is designed such that CMM effects are particularly large within the given constraints. Sets 2 and 3 are useful to study the dependence of CMM effects on the gluino mass and on the ratio $m_{\tilde{g}}/m_{\tilde{d}}$ separately, although always implying a change in the mass splitting $\Delta_{\tilde{d}}$.

2.5 The CP phase in $B_s - \overline{B}_s$ meson mixing

The characteristic effects of large atmospheric neutrino mixing in $b-s$ transitions introduce an additional source of CP violation to the Standard-Model CKM phase, which is the CMM phase ϕ_{B_s}, cf. Eq. (2.15). From the meson mixing element $(M_{12}^s)^{\text{CMM}}$ in Eq. (2.51) one learns that ϕ_{B_s} affects both the mass difference ΔM_s and the CP phase ϕ_s in the $B_s - \overline{B}_s$ system, defined in Eqs. (2.30) and (2.38), respectively. The recent measurements of these observables leave room for new-physics contributions. They compare to the Standard-Model predictions as follows,

$$\Delta M_s^{\text{exp}} = (117.0 \pm 0.8) \cdot 10^{-10} \text{ MeV} \qquad \phi_s^{\text{exp}} = (-0.77^{+0.29}_{-0.37}) \cup (-2.36^{+0.37}_{-0.29}) \text{ rad}$$
$$\Delta M_s^{\text{SM}} = (122.3^{+38}_{-31}) \cdot 10^{-10} \text{ MeV} \qquad \phi_s^{\text{SM}} = (-0.04 \pm 0.01) \text{ rad},$$
(2.55)

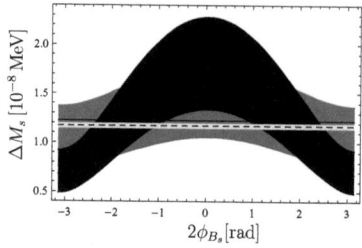 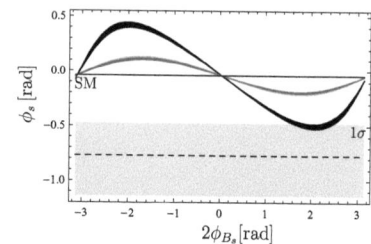

Figure 2.8: Effects of CMM phase ϕ_{B_s} in the $B_s - \bar{B}_s$ system for Set 1 (black curve), Set 2 (dark gray curve), and Set 3 (light gray curve), cf. Tab. 2.4, in comparison to the SM (black line). Left: ΔM_s, gray band: exp. 3-sigma range. Right: $\phi_s = \arg(-M_{12}/\Gamma_{12})$, gray band: exp. 1-sigma range, cf. Tab. 2.1. Sets 2 and 3 are largely superposed.

where the uncertainties for ΔM_s^{SM} and ϕ_s^{SM} are determined by varying the input values from Tab. 2.2 within their errors. The measurement of ΔM_s agrees with the SM expectation within the errors, which are dominated by hadronic uncertainties from F_{B_s} and $\hat{B}_{B_s}$. The best fit of the CP phase ϕ_s agrees with the SM central value at the level of 2.2σ only [63, 77]. This tension is often referred to as a hint for new physics in the B_s sector.
Let us first concentrate on ΔM_s. The measurement being precise and theoretical uncertainties under control, the mass difference in $B_s - \bar{B}_s$ sets an upper bound on CMM effects, as illustrated in Fig. 2.7 for $\phi_{B_s} = 0$. In this setup the lower bound on the gluino mass is strengthened to $m_{\tilde{g}} \gtrsim 550\,\text{GeV}$ at 3σ. This excludes $\phi_{B_s} = 0$ in our Set 1 where $m_{\tilde{g}} = 400\,\text{GeV}$. We thus turn the question around and ask for the constraints on the a priori free phase ϕ_{B_s} in a parameter set where CMM effects are sizeable. Fig. 2.8 left shows ΔM_s as a function of $2\phi_{B_s}$ for the three parameter sets from Tab. 2.4. Only in Set 1 the sensitivity to CMM effects is sufficient to derive constraints on the CMM phase (black band),

$$\Delta M_s, \text{ Set 1}: \quad 1.0 \lesssim |2\phi_{B_s}| \lesssim 2.4\,. \tag{2.56}$$

How large can CMM effects in the CP phase ϕ_s be within these bounds? In Fig. 2.8 right we plot $\phi_s(2\phi_{B_s})$ for the same CMM parameter sets. The sensitivity of ϕ_s to ϕ_{B_s} is increased with respect to ΔM_s. In Set 1, the experimental one-sigma band is reached for $2\phi_{B_s} \approx 2$, in agreement with the parameter region favoured by ΔM_s. Set 2 (dark gray curve) and Set 3 (light gray curve) largely coincide in their sensitivity to ϕ_{B_s}, indicating that the size of CMM effects in both ΔM_s and ϕ_s is dominated by the gluino mass $m_{\tilde{g}}$. We conclude that neutrino mixing within the CMM model generates a sizeable CP phase in $B_s - \bar{B}_s$ mixing for suitable SUSY parameters and CMM phase. Thereby it is possible, though difficult, to cure the tension between the measurement of ϕ_s and its Standard-Model prediction.

Chapter 3

Yukawa Corrections for Light Fermions

The measured masses of the bottom quark and the tau lepton are in line with bottom-tau unification at the GUT scale. For light fermions, however, down-quark-lepton unification has to be corrected by additional flavour structures in the Yukawa sector of a unified theory. Higher-dimensional operators, being suppressed by a large mass scale, provide small corrections for light fermions, while generally preserving bottom-tau unification. They give rise to new FCNC involving light right-handed down (s)quarks, such that the large atmospheric neutrino mixing angle also influences $s-d$ and $b-d$ transitions. The a priori arbitrary flavour structure of Yukawa corrections is phenomenologically strongly constrained from kaon physics, namely ϵ_K, and to lesser extent as well from B_d observables. Consequently, the flavour structure of Yukawa corrections must be basically aligned with the initial, unified Yukawa couplings. Within these strong constraints, we discuss the effects of Yukawa corrections in K and B physics on the unitarity triangle. We find that the effects of large neutrino mixing can remove a present tension concerning CP violation in the $K-\overline{K}$ and $B_d-\overline{B}_d$ meson systems.

3.1 Corrections from higher-dimensional operators

Down-quark-lepton Yukawa unification is a feature of SU(5) symmetry, where SU(2)-singlet down quarks are embedded together with the lepton doublet in a $\bar{5}$ representation. At the unification scale M_{GUT} the Yukawa couplings are thus equal up to transposition,

$$Y_d = Y_e^\top. \tag{3.1}$$

The renormalization group evolution of this relation down to the scale of electroweak symmetry breaking yields the correct masses of bottom quarks and tau leptons. For the light fermion generations, however, down-quark-lepton unification fails. Exact unification would predict the mass relation $m_s/m_d = m_\mu/m_e$, which is not fulfilled by the observed

ratios
$$\frac{m_s}{m_d} \approx 20, \qquad \frac{m_\mu}{m_e} \approx 200. \tag{3.2}$$

In SU(5), this discrepancy can be corrected by adding Yukawa couplings of mass dimension five without enlarging the content of Higgs fields [78]. The adjoint Higgs representation $\Sigma = 24_H$, which is needed for SU(5) $\to G_{\rm SM}$ breaking anyway, is suited to extend the Yukawa sector to

$$W^{d,e}_{\rm SU(5)} = Y^{ij}_{\bar 5} \, 10^{ab}_i \bar 5_{ja} \bar 5_{Hb} + (\widetilde Y'_{24})^{ij} \, 10^{ab}_i \bar 5_{ja} \frac{\Sigma^c_b}{M_{\rm Pl}} \bar 5_{Hc} + (\widetilde Y_{24})^{ij} \, 10^{ab}_i \frac{\Sigma^c_b}{M_{\rm Pl}} \bar 5_{jc} \bar 5_{Ha}. \tag{3.3}$$

Besides the dimension-four term of minimal SU(5), two additional independent Yukawa couplings $\widetilde Y'_{24}$ and $\widetilde Y_{24}$ are introduced by dimension-five terms. Once SU(5) is broken to the Standard Model via a vev $\langle \Sigma \rangle = v_5 \mathrm{diag}(2,2,2;-3,-3)$ with $v_5 = \mathcal{O}(M_{\rm GUT})$, the effective down and lepton Yukawa couplings read

$$\begin{aligned} Y_d &= Y_{\bar 5} - 3 \frac{v_5}{M_{\rm Pl}} \widetilde Y'_{24} + 2 \frac{v_5}{M_{\rm Pl}} \widetilde Y_{24}, \\ Y_e^\top &= Y_{\bar 5} - 3 \frac{v_5}{M_{\rm Pl}} \widetilde Y'_{24} - 3 \frac{v_5}{M_{\rm Pl}} \widetilde Y_{24}. \end{aligned} \tag{3.4}$$

Note that dimension-five terms $\sim \widetilde Y'_{24}$ merely shift both Yukawa couplings, without affecting their unification. These contributions, however, are important to suppress proton decay [79, 80]. The second coupling $\widetilde Y_{24}$ abrogates down-quark-lepton unification, because the Higgs field Σ is "sandwiched" between the fermions, leading to

$$Y_d = Y_e^\top + 5 \frac{v_5}{M_{\rm Pl}} \widetilde Y_{24}. \tag{3.5}$$

Since the correction $\widetilde Y_{24}$ is suppressed by $M_{\rm GUT}/M_{\rm Pl}$, it affects only the small Yukawa couplings of the first and second fermion generations, while preserving the successful bottom-tau unification.

In the CMM model, Yukawa corrections due to SU(5) breaking generally arise from the dimension-five term $\sim 45_H \otimes 10'_H$ in the SO(10) superpotential, cf. Eq. (2.3),

$$W^{d,e}_{\rm CMM} = 16_i \, Y^{ij}_{45} \, 16_j \, \frac{45_H \, 10'_H}{M_{\rm Pl}}. \tag{3.6}$$

This term generates both unified down and lepton Yukawa couplings, as well as corrections for the light generations. The embedding of the relevant SU(5) dimension-five couplings from Eq. (3.3) is made explicit by writing the superpotential after SO(10) breaking via a vev in the SU(5) singlet in 45_H, $\langle 1_{45} \rangle = v_{10}$,

$$W^{d,e}_{\rm CMM} = \frac{v_{10}}{M_{\rm Pl}} Y_{45} \, 10^{ab}_i \bar 5_{ja} \bar 5'_{Hb} + \widetilde Y'_{45} \, 10^{ab}_i \bar 5_{ja} \frac{\Sigma^c_b}{M_{\rm Pl}} \bar 5'_{Hc} + \widetilde Y_{45} \, 10^{ab}_i \frac{\Sigma^c_b}{M_{\rm Pl}} \bar 5_{jc} \bar 5'_{Ha}. \tag{3.7}$$

3.1. Corrections from higher-dimensional operators

Below the GUT scale, these operators yield the familiar down-quark-lepton Yukawa relation

$$Y_d = Y_e^\top + 5\frac{v_5}{M_{\text{Pl}}}\widetilde{Y}_{45}. \tag{3.8}$$

Naturally, one assumes that the entries of Y_{45} and $\widetilde{Y}_{45}$ have the same magnitude as $(Y_{10})^{33}$, the top Yukawa coupling. Then the corrections from $\widetilde{Y}_{45}$ are suppressed by $v_5/v_{10} \approx 10^{-1}$ with respect to the bottom and tau couplings, which are of $\mathcal{O}(v_{10}/M_{\text{Pl}}) y_t = 10^{-2} y_t$. Therefore effects of Yukawa corrections among light fermions are typically of $\mathcal{O}(1)$. Generally, the flavour structure of $\widetilde{Y}_{45}$ differs from Y_{45}, since there are four possible SO(10) structures for the coupling $(16_i 16_j)(45_H 10'_H)$, given in Eqs. (1.69a) – (1.69d). Concretely, Yukawa corrections à la $\widetilde{Y}_{45}$ can be generated by effective 120, 16, and/or 144 couplings at SO(10) level. The effective 10 coupling may in addition contribute equally to both Y_d and $Y_e^\top$ via Y_{45}. In presence of Yukawa corrections, the matrices Y_d and Y_e can no longer be diagonalized by the same rotation $V_q^* \widehat{Y}_{d,e} V_\ell$ as in Eq. (2.11). In the basis where Y_e is diagonal, we rather have

$$U_L \widehat{Y}_d U_R = \widehat{Y}_e + 5\frac{v_5}{M_{\text{Pl}}}\widetilde{Y}_{45}, \tag{3.9}$$

where the unitary matrices U_L and U_R account for the mismatch between Y_d and Y_e due to corrections from $\widetilde{Y}_{45}$. Since these corrections preserve bottom-tau unification, U_L and U_R have a non-trivial 1-2 block only,

$$U_L, U_R \sim \begin{pmatrix} * & * & 0 \\ * & * & 0 \\ 0 & 0 & 1 \end{pmatrix}. \tag{3.10}$$

The rotation matrix R_d of right-handed down (s)quarks is no longer directly linked to the PMNS matrix, but contains the additional rotations among light down (s)quarks $U \equiv U_R$,

$$R_d = (UV_\ell)^\top. \tag{3.11}$$

We parametrize the unitary matrix U by one rotation angle θ and four complex phases,

$$U = \begin{pmatrix} U_{11} & U_{12} & 0 \\ U_{21} & U_{22} & 0 \\ 0 & 0 & e^{i\phi_4} \end{pmatrix} = \begin{pmatrix} \cos\theta\, e^{i\phi_1} & -\sin\theta\, e^{i(\phi_1-\phi_2+\phi_3)} & 0 \\ \sin\theta\, e^{i\phi_2} & \cos\theta\, e^{i\phi_3} & 0 \\ 0 & 0 & e^{i\phi_4} \end{pmatrix}. \tag{3.12}$$

Clearly, in absence of Yukawa corrections $U = \mathbb{1}$, and we are back in the original CMM setup discussed in Chapter 2. Assuming tri-bi-maximal neutrino mixing, the down-(s)quark rotation matrix reads, with V_ℓ from Eq. (1.38) and $\theta_{13} = 0$,

$$R_d = \frac{1}{\sqrt{6}}\begin{pmatrix} 2U_{11}e^{i\alpha_1} - U_{12}e^{i\alpha_4} & 2U_{21}e^{i\alpha_1} - U_{22}e^{i\alpha_4} & e^{i(\phi_4+\alpha_5)} \\ \sqrt{2}e^{i\alpha_2}(U_{11}+U_{12}e^{i(\alpha_4-\alpha_1)}) & \sqrt{2}e^{i\alpha_2}(U_{21}+U_{22}e^{i(\alpha_4-\alpha_1)}) & -\sqrt{2}e^{i(\phi_4-\alpha_1+\alpha_2+\alpha_5)} \\ \sqrt{3}U_{12}e^{i(-\alpha_1+\alpha_3+\alpha_4)} & \sqrt{3}U_{22}e^{i(-\alpha_1+\alpha_3+\alpha_4)} & \sqrt{3}e^{i(\phi_4-\alpha_1+\alpha_3+\alpha_5)} \end{pmatrix}. \tag{3.13}$$

What are the physical implications of the additional rotations U on down-squark currents? In the super-CKM basis, where down-type Yukawa couplings are diagonal, the mass matrix of right-handed down squarks is given by

$$(M_{\tilde{d}}^2)^{\text{sCKM}} = R_d^\dagger (M_{\tilde{d}}^2)^U R_d$$
$$= m_{\tilde{d}}^2 \begin{pmatrix} 1 - \sin^2\theta \, \Delta_{\tilde{d}}/2 & \sin(2\theta) \, e^{-i\phi_K} \Delta_{\tilde{d}}/4 & \sin\theta \, e^{-i\phi_{B_d}} \Delta_{\tilde{d}}/2 \\ \sin(2\theta) \, e^{i\phi_K} \Delta_{\tilde{d}}/4 & 1 - \cos^2\theta \, \Delta_{\tilde{d}}/2 & -\cos\theta \, e^{-i\phi_{B_s}} \Delta_{\tilde{d}}/2 \\ \sin\theta \, e^{i\phi_{B_d}} \Delta_{\tilde{d}}/2 & -\cos\theta \, e^{i\phi_{B_s}} \Delta_{\tilde{d}}/2 & 1 - \Delta_{\tilde{d}}/2 \end{pmatrix}, \quad (3.14)$$

$$\phi_K = \phi_1 - \phi_2, \quad \phi_{B_s} = \phi_3 - \phi_4 + \alpha_4 - \alpha_5, \quad \phi_{B_d} = \phi_1 - \phi_2 + \phi_3 - \phi_4 + \alpha_4 - \alpha_5.$$

Comparing with Eq. (2.15), one observes that $M_{\tilde{d}}^2$ exhibits additional off-diagonal elements in the 1-2 and 1-3 sectors, resulting in $\tilde{d}_R - \tilde{s}_R$ and $\tilde{d}_R - \tilde{b}_R$ squark FCNC. The large atmospheric neutrino mixing is thus translated into sizeable effects in K and B_d physics, quantified by the rotation angle θ. Solar neutrino mixing is not observable, because the squark masses of the first two generations are degenerate. CMM effects in B_s observables are slightly reduced if $\theta \neq 0$. In summary, effects from atmospheric neutrino mixing in $b-s$ transitions generally imply contributions to $s-d$ and $b-d$ transitions if one corrects down-quark-lepton unification by additional flavour structures. Further, the new phases ϕ_K, ϕ_{B_d}, and ϕ_{B_s} induce CP violation beyond the Standard Model in all three down-squark FCNC. Note that these phases are not independent, but related by $\phi_{B_d} = \phi_{B_s} + \phi_K$.

Arguably, the predictivity for fermion masses and mixings strongly decreases with the new parameters introduced by Yukawa corrections. Yet the CMM model is not designed to predict the parameters related to Standard-Model particles, but to study the implications of down-quark-lepton unification in the mixing of their superpartners. The strong experimental constraints on FCNC in K and B_d observables can be used to gain insight into the flavour structure of Yukawa corrections in Grand Unified models. Concretely, these constraints translate into a strong upper limit on the angle θ. The CMM phases ϕ_K, ϕ_{B_d}, and ϕ_{B_s} are less constrained and can be adapted in order to generate considerable new-physics effects in CP-violating observables. In the following, we will derive constraints on θ from K and B meson mixing, in particular from the CP-violating quantity ϵ_K, and from the mass differences and CP asymmetries introduced in Sec. 2.3. Due to the additional entries in the down-squark mass matrix in Eq. (3.14), the mixing elements M_{12} in all three meson systems $K - \overline{K}$, $B_d - \overline{B}_d$, and $B_s - \overline{B}_s$ receive CMM contributions. They are characterized by the following combinations of rotation matrices in the relevant Wilson coefficients from Eq. (2.48),

$$M_{12}^K : [(R_d)_{32}(R_d)_{31}^*]^2, \quad M_{12}^d : [(R_d)_{33}(R_d)_{31}^*]^2, \quad M_{12}^s : [(R_d)_{33}(R_d)_{32}^*]^2. \quad (3.15)$$

By inserting the R_d elements using Eqs. (3.13) and (3.12), one arrives at explicit expressions for CMM contributions in meson mixing,

$$(M_{12}^K)^{\text{CMM}} = \frac{\alpha_s^2(M_Z)}{6m_{\tilde{g}}^2} \left[\frac{1}{4}\sin(2\theta)\, e^{-i\phi_K}\right]^2 M_K F_K^2 \widehat{B}_K \frac{\eta_2}{r} S^{(\tilde{g})}(r_1, r_3),$$

$$(M_{12}^d)^{\text{CMM}} = \frac{\alpha_s^2(M_Z)}{6m_{\tilde{g}}^2} \left[\frac{1}{2}\sin\theta\, e^{-i\phi_{B_d}}\right]^2 M_{B_d} F_{B_d}^2 \widehat{B}_{B_d} \frac{\eta_B}{r} S^{(\tilde{g})}(r_1, r_3), \quad (3.16)$$

$$(M_{12}^s)^{\text{CMM}} = \frac{\alpha_s^2(M_Z)}{6m_{\tilde{g}}^2} \left[\frac{1}{2}\cos\theta\, e^{-i\phi_{B_s}}\right]^2 M_{B_s} F_{B_s}^2 \widehat{B}_{B_s} \frac{\eta_B}{r} S^{(\tilde{g})}(r_1, r_3),$$

with all relevant quantities defined in Sec. 2.3.

3.2 Constraints from ϵ_K

The observable ϵ_K measures CP violation in $K - \overline{K}$ meson mixing. Since ϵ_K is a tiny quantity and in addition very accurately measured, it sets strong limits on CP-violating contributions to $s - d$ transitions. In the CMM model, ϵ_K constrains a combination of the parameters θ and ϕ_K, cf. Eq. (3.16), which originate from corrections to down-quark-lepton Yukawa unification for light fermions. These constraints hint at the flavour structure of higher-dimensional Yukawa terms, namely $\tilde{Y}_{45}$ in Eq. (3.9).
Experimental access to ϵ_K is given by the measurement of the amplitudes for K decays into two-pion final states $\langle \pi\pi | H | K \rangle$,

$$\epsilon_K \equiv \frac{\eta_{00} + 2\,\eta_{+-}}{3}, \qquad \eta_{00} = \frac{\langle \pi^0\pi^0 | H | K_L \rangle}{\langle \pi^0\pi^0 | H | K_S \rangle}, \qquad \eta_{+-} = \frac{\langle \pi^+\pi^- | H | K_L \rangle}{\langle \pi^+\pi^- | H | K_S \rangle}. \quad (3.17)$$

The final states $(\pi\pi)_I$ are classified in terms of their isospin quantum number $I = 0, 2$. Due to the suppression of the $I = 2$ decay amplitude with respect to $I = 0$, $A_2/A_0 \approx 1/22$, ϵ_K is well approximated by

$$\epsilon_K = \frac{\langle (\pi\pi)_{I=0} | K_L \rangle}{\langle (\pi\pi)_{I=0} | K_S \rangle} + \mathcal{O}\left(\frac{A_2^2}{A_0^2}\right). \quad (3.18)$$

By making use of empirical relations within the $K - \overline{K}$ system, one can express ϵ_K in terms of the mixing element M_{12}^K [67],

$$\epsilon_K = e^{i\phi_\epsilon} \sin\phi_\epsilon \left(\frac{\text{Im}M_{12}^K}{\Delta M_K} + \arg A_0\right). \quad (3.19)$$

The phase ϕ_ϵ in ϵ_K is measured to be [54]

$$\phi_\epsilon = \arctan\frac{2\Delta M_K}{\Delta\Gamma_K} = (43.51 \pm 0.05)^\circ. \quad (3.20)$$

3. Yukawa Corrections for Light Fermions

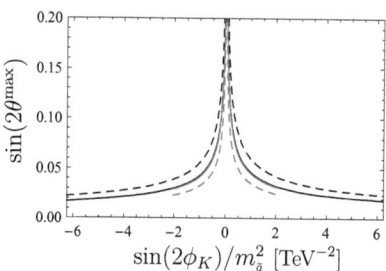

Figure 3.1: Constraints on θ from $|\epsilon_K|$ for Set 1 (black) and Set 2 (gray) (curves lie upon another). The dashed lines show the results for interchanged mass ratio, $m_{\tilde{d}}/m_{\tilde{g}} = 2.86$ (black dashed) and 5 (gray dashed).

The deviation of ϕ_ϵ from $\pi/4$ and the estimation of $\mathcal{O}(5\%)$ contributions from the phase of the isospin-zero amplitude A_0 are factored out in the quantity $\kappa_\epsilon = (0.92 \pm 0.02)$ [67]. This leads to the compact formula we will use to constrain our CMM parameters,

$$|\epsilon_K| = \kappa_\epsilon \frac{\mathrm{Im} M_{12}^K}{\sqrt{2}\Delta M_K}. \tag{3.21}$$

Comparing with Eq. (3.16), we find the CMM contribution to $|\epsilon_K|$ proportional to $\mathrm{Im}(M_{12}^K)^{\mathrm{CMM}} \sim \sin^2(2\theta)\sin(2\phi_K)/m_{\tilde{g}}^2$. If the phase ϕ_K does not vanish and the gluino mass is not too large, one generally expects strong constraints on θ for the following reasons:

- In the Standard Model, ϵ_K is twofold suppressed: The GIM mechanism predicts the weak W box diagrams to be proportional to m_q^2/m_W^2, which suppresses contributions with $q = u, c$ quarks in the loop. Second, the a priori large top quark contribution is strongly CKM-suppressed by $(V_{td}^* V_{ts})^2$, cf. Eq. (2.45).

- On the contrary, the main CMM contributions stem from gluino box diagrams based on strong interactions, which involve a larger coupling constant and avoid CKM suppression.

- The Standard-Model prediction for ϵ_K agrees with the precise measurement, leaving only little space for new physics.

The first two arguments apply likewise to the mass difference ΔM_K. This quantity, however, is plagued by short-distance contributions, which are difficult to estimate and result in large theoretical uncertainties. ϵ_K is thus the most effective observable to constrain θ. In Fig. 3.1, we show the limits on the $\tilde{s}_R - \tilde{d}_R$ mixing angle θ from $|\epsilon_K|$ dependent on the relevant combinations of parameters, which are $\sin(2\phi_K)/m_{\tilde{g}}^2$, $m_{\tilde{d}}/m_{\tilde{g}}$, and $\Delta_{\tilde{d}}$. The

3.3. Constraints from $B_d - \bar{B}_d$ and $B_s - \bar{B}_s$ mixing

	$m_{\tilde{g}}$ [GeV]	$m_{\tilde{d}}$ [GeV]	$\Delta_{\tilde{d}}$	$\theta^{\max}$ [°]
Set 1	400	2000	0.52	0.5
Set 2	700	2000	0.44	0.9
Set 3	700	3000	0.51	0.9

Table 3.1: CMM parameter sets for fixed $a_d/m_{\tilde{d}} = 1.8$, $\arg(\mu) = 0$, and $\tan\beta = 5$. The last column shows the maximal $\tilde{s}_R - \tilde{d}_R$ mixing angle $\theta^{\max}$ allowed by $|\epsilon_K|$ for $\sin(2\phi_K) = 1$ (the symmetric solution $\theta \in [\pi/2 - \theta^{\max}, \pi/2]$ is excluded by B physics observables, see Sec. 3.3).

black and gray lines correspond to the CMM parameter sets 1 and 2 introduced in Sec. 2.4, respectively, while the dashed lines are obtained by interchanging $m_{\tilde{d}}/m_{\tilde{g}}$. The dependence on $m_{\tilde{d}}/m_{\tilde{g}}$ and $\Delta_{\tilde{d}}$ is mild, which can be seen by comparing the plain and dashed lines of the same $(m_{\tilde{d}}/m_{\tilde{g}})$ and interchanged $(\Delta_{\tilde{d}})$ gray level. Typically, for $|\sin(2\phi_K)|/m_{\tilde{g}}^2 \gtrsim 1\,\text{TeV}^{-2}$, $\theta^{\max}$ is of the order of one degree. Fig. 3.1 has been obtained treating the errors of the input parameters in Tab. 2.2 as flat, yet a different error treatment would not change this picture significantly. Fixing ϕ_K to $\pi/4$, the precise limits on θ obtained for the three parameter sets defined in Sec. 2.4 are displayed in the last column of Tab. 3.1.

Constraints from ΔM_K are relevant only if ϕ_K is close to zero. Due to the large hadronic uncertainties, we impose $2|(M_{12}^K)^{\text{CMM}}| < \Delta M_K^{\exp}$ to stay on the conservative side. In this case, for $m_{\tilde{g}} \simeq 700\,\text{GeV}$, the constraint from ΔM_K only starts to compete with that from $|\epsilon_K|$ when $|\phi_K| = \mathcal{O}(0.1°)$, corresponding to $\theta^{\max} \simeq 10° - 20°$ (depending on the precise values of $\Delta_{\tilde{d}}$ and $m_{\tilde{d}}/m_{\tilde{g}}$). In Set 1, 2, and 3, ΔM_K puts the constraints $\theta^{\max} = 10°$, $19°$, and $18°$, respectively.

Finally, we briefly comment on the dependence of $\theta^{\max}$ on the hypothesis of tri-bi-maximal lepton mixing. In particular, one might expect the 2-3 mixing angle to be large but not exactly equal to $\pi/4$. In this case, $\text{Im}\left[(R_d)_{32}(R_d)_{31}^*\right]^2 = -\frac{1}{4}\sin^4\theta_{23}\sin^2(2\theta)\sin(2\phi_K)$ for $\theta_{13} = 0$. Hence, for large θ_{23}, the constraints on θ do not differ much. For a sizeable 1-3 mixing angle in V_ℓ, $|\epsilon_K|$ gets additional contributions,

$$\Delta\left(\text{Im}\left[(R_d)_{32}(R_d)_{31}^*\right]^2\right) = \sin\theta_{13}\sin^3\theta_{23}\sin(2\theta)\left[-\cos(2\phi_K)\sin(\phi_3 - \phi_2 + \alpha_4 - \alpha_1 - \delta)\right.$$
$$\left. + \sin(2\phi_K)\cos(2\theta)\cos(\phi_3 - \phi_2 + \alpha_4 - \alpha_1 - \delta)\right] + \mathcal{O}\left(\sin^2\theta_{13}\right). \quad (3.22)$$

No large numerical factors offset the $\sin\theta_{13}$-suppression, such that the modified bounds on θ are again as stringent as those exemplified in Fig. 3.1.

3.3 Constraints from $B_d - \bar{B}_d$ and $B_s - \bar{B}_s$ mixing

The observables related to $B - \bar{B}$ mixing are generally less sensitive to CMM contributions, mainly because the SM contributions from weak box diagrams are not as strongly CKM-suppressed as in the kaon system. For $\phi_K \simeq 0\,(\pi/2)$ however, where CMM effects disappear from $|\epsilon_K|$, B physics observables provide useful constraints on the parameters θ and

$\phi_{B_s} = \phi_{B_d}(\phi_{B_d} - \pi/2)$. In Fig. 3.2, we show the parameter regions allowed by ΔM_d, $S_{J/\psi K_S}$, and $\Delta M_d/\Delta M_s$ for Set 1 (gray) and Set 2 (black) for $\phi_K = 0$ and $\phi_K = \pi/2$. Other values of ϕ_K lead qualitatively to the same results. Note that the constraint from $\Delta M_d/\Delta M_s$ depends on both ϕ_{B_s} and $\phi_{B_d} = \phi_K + \phi_{B_s}$. Further, ΔM_d does not give additional constraints with respect to the ratio $\Delta M_d/\Delta M_s$ for $\phi_K = 0$, which is due to the fact that the ratio of hadronic B_s and B_d parameters ξ has a smaller uncertainty than $F_{B_d}\hat{B}_{B_d}^{1/2}$, cf. Tab. 2.2. The combination of ΔM_d, $S_{J/\psi K_S}$, and $\Delta M_d/\Delta M_s$ generally constrains θ to be below $\theta^{\max} \simeq 10° - 20°$, similar to ΔM_K. The solution $\theta \in [\pi/2 - \theta^{\max}, \pi/2]$ left by $|\epsilon_K|$ and ΔM_K is thereby excluded.

Importantly, $\Delta M_d/\Delta M_s$, ΔM_s, and ϕ_s rigorously constrain the phase ϕ_{B_s} for small θ, as could already be seen in Fig. 2.8 for the latter two observables. The constraints from ΔM_s are included in Fig. 3.2. The B_s phase ϕ_s can cut further into the parameter space, especially for negative values of ϕ_{B_s}. In Fig. 3.2, ϕ_s would remove gray points with $2\phi_{B_s} < 0$ and black points with $-2.4 < 2\phi_{B_s} < -1$. Roughly speaking, $B - \bar{B}$ mixing observables favour positive values of ϕ_{B_s} in the CMM model with Yukawa corrections.

Finally, we can distinguish two scenarios of CMM effects in $B - \bar{B}$ mixing:

- $\phi_K \neq 0$: Effects in $|\epsilon_K|$ are large, constraining θ to be very small. In this case, there are no visible CMM effects in B_d observables like ΔM_d and $S_{J/\psi K_S}$. Contributions to B_s observables are close to maximal.

- $\phi_K \simeq 0$: The CMM contributions to $|\epsilon_K|$ vanish, but θ is (less) constrained from $B - \bar{B}$ mixing and ΔM_K. CMM effects are thus sizeable in B_d observables and only slightly lowered in the B_s system. The CMM phases in B_d and B_s observables are equal, $\phi_{B_s} \simeq \phi_{B_d}$.

3.4 Closing the unitarity triangle

Recent works on CP violation in meson mixing pointed out a tension between the $K - \bar{K}$ and $B_d - \bar{B}_d$ systems: The amount of CP violation in $B_d - \bar{B}_d$ mixing, extracted from $S_{J/\psi K_S}$, seems to be inconsistent with CP violation in $K - \bar{K}$ mixing, measured by ϵ_K [67, 81–83]. One thus predicts new CP phases in the B_d and/or K sectors. We will show that, within the CMM model, it is possible to restore consistency by making use of CP effects in $s-d$, $b-d$, and $b-s$ transitions. These effects simultaneously generate a sizeable phase ϕ_s in the $B_s - \bar{B}_s$ system, which is welcome to explain the amount of CP violation measured in $B_s \to J/\psi\phi$ decays.

The tension between the $K - \bar{K}$ and $B_d - \bar{B}_d$ systems is made explicit by comparing the value of $\sin(2\beta)$ extracted from $S_{J/\psi K_S}$, cf. Eq. (2.35), to its determination from $|\epsilon_K|$, whose leading contribution is proportional to $\sin(2\beta)$, see Eq. (3.23) below. The angle β being one of four parameters that fix the unitarity triangle, and thereby quark flavour mixing, we choose the three further inputs as $|V_{us}|$, $|V_{cb}|$, and $\Delta M_s/\Delta M_d$. In the Standard Model, $|\epsilon_K|$ and $\Delta M_s/\Delta M_d$ can be expressed in terms of $\sin(2\beta)$ and a side of the unitarity

3.4. Closing the unitarity triangle

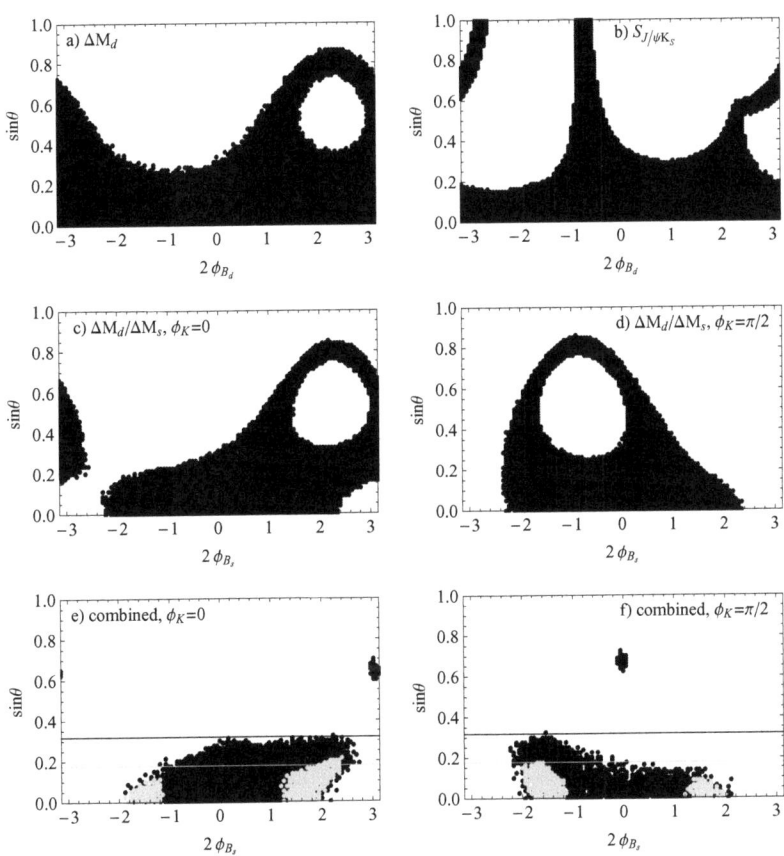

Figure 3.2: Constraints on θ from B physics observables. Black (gray) points indicate allowed regions in parameter space for Set 2 (Set 1). The first four plots show individual three-sigma constraints from (a) ΔM_d, (b) $S_{J/\psi K_S}$, (c) $\Delta M_d/\Delta M_s$ setting $\phi_K = 0$, (d) $\Delta M_d/\Delta M_s$ setting $\phi_K = \pi/2$. Plots (e) and (f) show the combined (a,b,c) and (a,b,d) constraints, respectively. In the case of Set 1, the three-sigma constraint from ΔM_s has been included, excluding points outside the range $1.3 \lesssim |2\phi_{B_s}| \lesssim 2.3$ (Set 2 is not affected by this constraint, cf. Fig. 2.8 left). ΔM_K excludes points above the black (gray) horizontal line in Set 2 (Set 1).

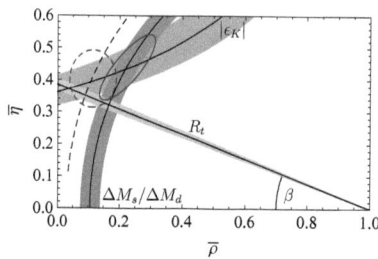
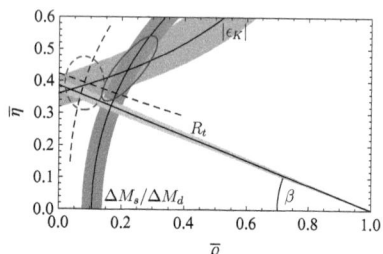

Figure 3.3: One-sigma constraints on the unitarity triangle from $S_{J/\psi K_S}$ (light gray), $|\epsilon_K|$ (gray), and $\Delta M_s/\Delta M_d$ (dark gray) in the SM. The one-sigma region determined from $|V_{us}|$, $|V_{cb}|^{\text{av}}$, $|\epsilon_K|$, and $\Delta M_s/\Delta M_d$ assuming the SM is shown in black, and its shift due to CMM effects is the dashed ellipse. Left: Scenario I, $\theta = 0, \phi_{B_s} = 0.7$. Right: Scenario II, $\theta = 0.1, \phi_{B_s} = \phi_{B_d} = 0.8$. CMM inputs: Set 1.

triangle $R_t = |V_{td} V_{tb}^*|/|V_{cd} V_{cb}^*|$ as

$$|\epsilon_K| = \kappa_\epsilon \frac{G_F^2 M_W^2}{12\pi^2} \frac{M_K F_K^2 \widehat{B}_K}{\sqrt{2} \Delta M_K} |V_{cb}|^2 |V_{us}|^2 \Big\{ |V_{cb}|^2 R_t^2 \sin(2\beta)\, \eta_2 S_0(x_t)$$
$$+ 2 R_t \sin\beta \left(\eta_3 S_0(x_c, x_t) - \eta_1 S_0(x_c) \right) \Big\}, \quad (3.23)$$

$$\frac{\Delta M_s}{\Delta M_d} \simeq \xi^2 \frac{M_{B_s}}{M_{B_d}} \frac{1}{R_t^2 |V_{us}|^2}.$$

Note that the determination of $\sin(2\beta)$ from the above expressions strongly depends on $|V_{cb}|$, because $|\epsilon_K| \sim |V_{cb}|^4$ in its leading contribution. Since the averaged value $|V_{cb}|^{\text{incl}}$ from inclusive semileptonic B decays in Tab. 2.2 is significantly larger than from exclusive $B \to D^* \ell \nu$ decays, $|V_{cb}|^{\text{excl}} = (38.8 \pm 1.1) \cdot 10^{-3}$ [63], it is instructive to compare the respective results for $\sin(2\beta)$ from $|\epsilon_K|$ with its determination from $S_{J/\psi K_S}$,

$$\sin(2\beta^d) = 0.671 \pm 0.024 \qquad S_{J/\psi K_S}$$
$$\sin(2\beta^K) = 0.81^{+0.11}_{-0.09} \qquad |\epsilon_K|, |V_{cb}|^{\text{incl}} \qquad (3.24)$$
$$= 0.98^{+0.02}_{-0.11} \qquad |\epsilon_K|, |V_{cb}|^{\text{excl}}.$$

The remaining input values are taken from Tab. 2.2. Note that with $|V_{cb}|^{\text{incl}}$ there is no significant deviation. With the smaller value $|V_{cb}|^{\text{excl}}$, $\sin(2\beta^K)$ is indeed larger than $\sin(2\beta^d)$, indicating an additional source of CP violation in the $B_d - \overline{B}_d$ system or compensating CP effects in $|\epsilon_K|$ and $\Delta M_s/\Delta M_d$. The hadronic corrections to $|\epsilon_K|$ from $\kappa_\epsilon = 0.92$ and a recent decrease of the bag parameter $\widehat{B}_K$ consolidate this discrepancy. In order to illustrate how CMM contributions can alleviate an observed tension, we will work with a weighted average of inclusive and exclusive determinations, $|V_{cb}|^{\text{av}} = (41.0 \pm 0.63) \cdot 10^{-3}$ [83]. In the CMM model with Yukawa corrections for light fermions, both $|\epsilon_K|$ and $\Delta M_s/\Delta M_d$ receive

3.4. Closing the unitarity triangle

additional CP-violating contributions. Thus $\sin(2\beta^K)$ is extracted from the more complex system of equations, which holds to 0.5% accuracy,

$$|\epsilon_K| = \kappa_\epsilon \frac{M_K F_K^2 \widehat{B}_K}{\sqrt{2}\Delta M_K} \times \left\{ \frac{G_F^2 M_W^2}{12\pi^2} |V_{cb}|^2 |V_{us}|^2 \left[|V_{cb}|^2 R_t^2 \sin(2\beta) \eta_2 S_0(x_t) \right.\right.$$

$$\left.\left. + 2R_t \sin\beta \left(\eta_3 S_0(x_c, x_t) - \eta_1 S_0(x_c)\right)\right] - \frac{\alpha_s^2(M_Z)}{6m_{\tilde{g}}^2} \frac{1}{16} \sin^2(2\theta)\sin(2\phi_K)\frac{\eta_2}{r} S^{(\tilde{g})}(r_1, r_3) \right\},$$

$$\frac{\Delta M_s}{\Delta M_d} = \xi^2 \frac{M_{B_s}}{M_{B_d}}$$

$$\times \frac{\left\{ (k_1 + X\cos^2\theta\cos 2\phi_{B_s})^2 + (-2k_2 R_t \sin\beta |V_{us}|^2 - X\cos^2\theta\sin 2\phi_{B_s})^2 \right\}^{1/2}}{\left\{ (R_t^2 \cos 2\beta |V_{us}|^2 + X\sin^2\theta\cos 2\phi_{B_d})^2 + (R_t^2 \sin 2\beta |V_{us}|^2 - X\sin^2\theta\sin 2\phi_{B_d})^2 \right\}^{1/2}},$$

with $k_1 = 1 + |V_{us}|^2(1 - 2R_t\cos\beta)$, $\qquad X = \dfrac{\pi^2 \alpha_s^2(M_Z) S^{(\tilde{g})}(r_1, r_3)}{2|V_{cb}|^2 G_F^2 M_W^2 m_{\tilde{g}}^2 r S_0(x_t)},$

$k_2 = 1 + |V_{us}|^2(1 - R_t\cos\beta)$. $\hfill (3.25)$

Due to the high sensitivity of $|\epsilon_K|$ to CMM effects, either θ or ϕ_K have to be very small. We therefore consider the two limiting scenarios $\theta = 0$ and $\phi_K = 0$ for which $|\epsilon_K|$ is not affected by CMM contributions and study the respective CP effects in $\Delta M_s/\Delta M_d$ and $S_{J/\psi K_S}$. We use the CMM input parameters of Set 1. All errors are treated as gaussian.

I) $\theta = 0$: CMM effects in $\Delta M_s/\Delta M_d$ Since for $\theta = 0$ there are no effects in $K - \overline{K}$ and $B_d - \overline{B}_d$ mixing, CMM contributions enter the unitarity triangle only via ΔM_s. From Fig. 3.3 left one sees that R_t has to increase in order to close the UT. This requires a CP-violating phase $2\phi_{B_s} \in [1.3, 1.8]$, taking into account the three-sigma constraints on ϕ_{B_s} from ΔM_s and ϕ_s. The dashed curve shows R_t for $\phi_{B_s} = 0.7$, such that the UT determined from $|\epsilon_K|$ and $\Delta M_s/\Delta M_d$ agrees with the $\sin(2\beta)$ measurement from $S_{J/\psi K_S}$.

II) $\phi_K = 0$, $\theta = 0.1$: CMM effects in $\Delta M_s/\Delta M_d$ and $S_{J/\psi K_S}$ In this second case, CMM contributions enter both $\Delta M_s/\Delta M_d$ and $S_{J/\psi K_S}$. For a fixed angle θ, the unitarity triangle can be closed by adjusting the CMM phase $\phi_{B_s} = \phi_{B_d}$. The resulting apex of the UT is shown by the intersection of the dashed lines in Fig. 3.3 right for $\theta = 0.1$ and $\phi_{B_s} = 0.8$. For any value of θ allowed by the constraints from B_d and B_s observables in Sec. 3.3 one can find a phase ϕ_{B_s} to close the UT.

Departures from the limits $\theta = 0$ and $\phi_K = 0$ rapidly generate significant additional effects in $|\epsilon_K|$, cf. Fig. 3.1. These contributions can lower the band from the $|\epsilon_K|$ constraint in the $(\bar{\rho}, \bar{\eta})$ plane, directly compensating for small values of $|V_{cb}|$ and $\widehat{B}_K$. Interestingly, the values of the CMM phase $2\phi_{B_s} \simeq 1.5$, favoured to close the unitarity triangle from K and B_d observables, simultaneously lead to a sizeable CP phase ϕ_s in the $B_s - \overline{B}_s$ system, cf. Fig. 2.8. We conclude that within the CMM model, despite the strong constraints on

$s-d$ and $b-d$ FCNC, CP-violating effects are sufficiently large to consistently explain the observed SM tensions concerning CP violation in $K-\overline{K}$, $B_d-\overline{B}_d$, and $B_s-\overline{B}_s$ mixing.

3.5 The flavour structure of Yukawa corrections

The probes of Yukawa corrections in $s-d$ and $b-d$ transitions lead to strong constraints on the angle θ, which parametrizes deviations from the unification of light fermions. For a non-vanishing phase ϕ_K, $|\epsilon_K|$ provides the upper bound $\theta^{\max} = \mathcal{O}(0.1°)$. The B_d observables ΔM_d, $S_{J/\psi K_S}$, and $\Delta M_d/\Delta M_s$ lead to $\theta^{\max} = \mathcal{O}(20°)$, independently from the respective CMM phase constellations. What are the implications of these constraints on the flavour structure of dimension-five Yukawa corrections?
The constraints from $|\epsilon_K|$ require the correction matrix $U = U_R$ to be essentially diagonal, specifying the corrected down-quark-lepton relation from Eq. (3.9) to

$$Y_d = U_L \widehat{Y}_d \widehat{U}_R = \widehat{Y}_e + 5\frac{v_5}{M_{\text{Pl}}}\widetilde{Y}_{45}. \qquad (3.26)$$

Since U_R diagonalizes $Y_d^\dagger Y_d$, in the basis where Y_e is diagonal the combination $\widehat{Y}_e \widetilde{Y}_{45} + \widetilde{Y}_{45}^\dagger \widehat{Y}_e + 5\frac{v_5}{M_{\text{Pl}}}\widetilde{Y}_{45}^\dagger \widetilde{Y}_{45}$ must be diagonal as well. Without miraculous cancellations, this implies that *the Yukawa corrections $\widetilde{Y}_{45}$ must be aligned with the couplings Y_d and $Y_e^\top$*. Going back to Eq. (3.26), we see that U_L, the matrix which governs FCNC with light left-handed sleptons, has to be diagonal as well, i.e.

$$U_L, U_R \sim \begin{pmatrix} * & 0 & 0 \\ 0 & * & 0 \\ 0 & 0 & 1 \end{pmatrix}. \qquad (3.27)$$

Once more the GUT relation between quarks and leptons strikes: The study of Yukawa corrections in the down-quark sector implies that FCNC involving $\tilde{e}_L$ are very small if the constraint from $|\epsilon_K|$ is rigorous. For $\phi_K = 0$, from B_d constraints one still derives that flavour-violating $\tilde{\mu}_L - \tilde{e}_L$ and $\tilde{\tau}_L - \tilde{e}_L$ transitions are naturally not larger than $\tilde{s}_R - \tilde{d}_R$ and $\tilde{b}_R - \tilde{d}_R$ transitions.

While we have worked out the analysis for a specific SO(10) model, these results hold in general for GUT models with small Higgs representations. An efficient mechanism is generally needed to render neutrino mixing in right-handed down-squark FCNC visible. In the CMM model, this mechanism is provided by the fast SO(10) running of the $\tilde{d}_R$ soft mass matrix, which generates the large universality-breaking quantity $\Delta_{\tilde{d}}$ at the electroweak scale. Of course, other GUT scenarios could include additional sources of flavour and CP violation leading to down-squark FCNC, like the SU(5) couplings of right-handed down (s)quarks to heavy neutrinos introduced in Sec. 1.3.[1] These effects could soften the

---

[1] In the CMM model, these effects are very small, because the neutrino Yukawa coupling Y_ν has the same strong hierarchy as Y_{10}, thus suppressing FCNC for light down (s)quarks.

3.5. The flavour structure of Yukawa corrections

constraints on θ. Yet, they would have to be fairly fine-tuned to cancel the potentially large Yukawa corrections from SU(5)-breaking dimension-five terms.

All in all, corrections to down-quark-lepton Yukawa unification from higher-dimensional operators cannot introduce new flavour structures with respect to the initially unified couplings.

Chapter 4

Supersymmetric Unification and Large tan β

The unification of third-generation Yukawa couplings is highly sensitive to the parameter $\tan\beta$. Knowing $\tan\beta$ substantially helps to specify the Yukawa sector in supersymmetric SO(10) models. In a scenario with large $\tan\beta$, Higgs couplings to down-type fermions are enhanced, and loop-suppressed down-(s)quark FCNC can be large. The decay modes $B \to \tau\nu$ and $B \to D\tau\nu$ are sensitive to $\tan\beta/M_{H^+}$ at tree level. Differential distributions in $B \to D\tau\nu$ are particularly well suited to determine the magnitude and phase of the charged-Higgs coupling, and thereby to derive rigorous constraints on $\tan\beta$ for a fixed SUSY spectrum. The theoretical accuracy of $B \to D\tau\nu$ distributions is limited by hadronic form factors, which however are known with less than 7% uncertainty. Experimentally, these modes should be detectable at the B factories with current or upgraded statistics. The resulting constraints on $\tan\beta$ help to classify SO(10) models with respect to their Yukawa sector.

4.1 Effects of large tan β in flavour physics

The ratio of the two Higgs-doublet vevs, $\tan\beta = v_u/v_d$, is a free parameter in the Higgs sector of the MSSM and a key parameter in supersymmetric flavour physics. First, $\tan\beta$ governs the relative magnitude of top and bottom Yukawa couplings via the tree-level relation $m_t/m_b = y_t/y_b \cdot \tan\beta$. In the case of large $\tan\beta \approx 50$, the bottom Yukawa coupling is of the same order of magnitude as the otherwise dominant top Yukawa coupling. The effects of y_b in the RGE of couplings and squark masses are thus comparable with y_t, affecting the spectrum of SUSY particles and the conditions of Yukawa unification. We will come back to these issues in Sec. 4.4.

Second, the parameter β quantifies the mixing within the Higgs sector, cf. Eq. (1.24), and thereby enters Higgs-fermion couplings. In particular, the Yukawa couplings of charged Higgs bosons to right-handed down-type quarks can be enhanced by $\tan\beta$. This feature allows us to probe $\tan\beta$ in charged currents already at tree level. To compare the relative

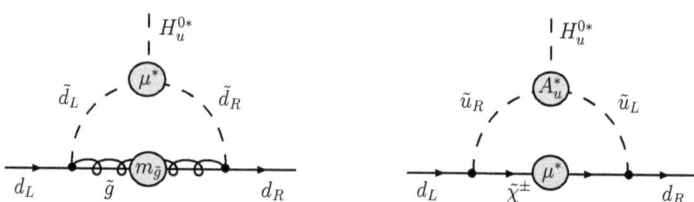

Figure 4.1: Contributions to non-holomorphic Yukawa couplings. Left: Down-quark-squark-gluino loop $\epsilon_d^{\tilde{g}}$. Right: Down-quark-up-squark-chargino loop $\epsilon_d^{\tilde{\chi}}$.

magnitude of H^+ and W^+ boson exchange, let us write down the MSSM Lagrangean for charged quark currents in terms of mass eigenstates, using Eqs. (1.27) and (1.24),

$$\mathcal{L}_q^{\text{CC}} = -\frac{g_2}{\sqrt{2}}\bar{d}_L V_{\text{CKM}}^\dagger \gamma^\mu u_L W_\mu^- + \bar{d}_R \widehat{Y}_d V_{\text{CKM}}^\dagger u_L \sin\beta H^- + \bar{d}_L \widehat{Y}_u V_{\text{CKM}}^\dagger u_R \cos\beta H^- + \text{h.c.}. \quad (4.1)$$

One observes that the coupling of charged Higgs bosons to right-handed down quarks is enhanced by $\tan\beta$ with respect to the coupling to left-handed down quarks, the latter being negligible if $\tan\beta$ is large. We further express the charged-Higgs coupling in terms of measured quantities and get

$$\mathcal{L}_q^{\text{CC}} = -\frac{g_2}{\sqrt{2}}\bar{d}_L V_{\text{CKM}}^\dagger \gamma^\mu u_L W_\mu^- + \frac{g_2}{\sqrt{2}}\frac{m_d}{M_W}\tan\beta\, \bar{d}_R V_{\text{CKM}}^\dagger u_L H^- + \text{h.c.}. \quad (4.2)$$

For $\tan\beta = \mathcal{O}(50)$, the contributions from charged-Higgs currents can thus compete with the longitudinal modes of W-boson exchange, leading to significant effects in helicity-suppressed meson decays. The most promising channels to find large-$\tan\beta$ effects in charged currents are the (semi)leptonic decays $B \to (D)\tau\nu$.

A third source of $\tan\beta$-enhanced effects in flavour physics arises from non-holomorphic Higgs-fermion couplings. Even though the superpotential must be holomorphic as in Eq. (1.20), such terms can be induced by supersymmetric loop corrections. They lead to effective couplings of fermions to the "wrong" Higgs doublet, which are $\tan\beta$-enhanced for down quarks and charged leptons [84, 85]. The effective Lagrangean for down-quark Yukawa couplings in the down mass eigenbasis is then given by

$$\mathcal{L}_d^{\text{eff}} = \bar{d}_R \widehat{Y}_d Q_L H_d + \bar{d}_R \widehat{Y}_d \epsilon_d Q_L H_u^* + \text{h.c.}, \qquad \epsilon_d = \epsilon_d^{\tilde{g}} + \epsilon_d^{\tilde{\chi}}. \quad (4.3)$$

The dominant contributions to non-holomorphic couplings stem from down-squark-gluino ($\epsilon_d^{\tilde{g}}$) and up-squark-chargino ($\epsilon_d^{\tilde{\chi}}$) loops, depicted in Fig. 4.1. The general structure of Yukawa corrections in the MSSM can be understood from symmetry considerations: Since fermion couplings to the "wrong" Higgs doublet break the holomorphy of the superpotential, the corresponding loop functions must be proportional to at least one power of

4.1. Effects of large tan β in flavour physics

a SUSY-breaking parameter. For the gluino loop this is the soft gauge boson mass $m_{\tilde{g}}$, for the chargino loop it is the trilinear coupling A_u. Besides breaking supersymmetry, non-holomorphic terms violate a global U(1) symmetry, known as Peccei-Quinn symmetry PQ [86, 87].[1] Under a Peccei-Quinn transformation, SU(2)-singlet down-type fields and the Higgs doublet H_d undergo a rephasing,

$$U(1)_{PQ}: \quad d_R \xrightarrow{PQ} e^{i\varphi} d_R, \quad e_R \xrightarrow{PQ} e^{i\varphi} e_R, \quad H_d \xrightarrow{PQ} e^{i\varphi} H_d. \qquad (4.4)$$

The MSSM superpotential preserves PQ, apart from the term $\mu H_u H_d$. Since non-holomorphic couplings $\bar{d}_R d_L H_u^*$ obviously violate PQ, they must be proportional to the higgsino mass parameter μ, cf. Fig. 4.1. Explicit expressions for the loop functions $\epsilon_d^{\tilde{g}}$ and $\epsilon_d^{\tilde{\chi}}$ are given in Ref. [88]. These contributions change the relation between the down-quark Yukawa coupling and the physical mass. By writing the effective Lagrangean in its vacuum state,

$$\mathcal{L}_d^{\text{eff}} = \bar{d}_R \widehat{Y}_d Q_L v \cos\beta + \bar{d}_R \widehat{Y}_d \epsilon_d Q_L v \sin\beta + \text{h.c.}, \qquad (4.5)$$

one derives the corrected relation

$$m_d = y_d v_d \left(1 + \epsilon_d \tan\beta\right). \qquad (4.6)$$

Since the loop suppression of $1/16\pi^2$ in ϵ_d is compensated by a large factor $\tan\beta$, supersymmetric loop corrections to the down-quark mass can be substantial. Note that this formula resums $\tan\beta$-enhanced radiative corrections to all orders in perturbation theory [89]. Apart from contributing to mass renormalization, non-holomorphic terms also induce flavour-changing couplings among down-type quarks. In the scenario of minimal flavour violation, these arise from chargino loops $\epsilon_d^{\tilde{\chi}}$ only, which add flavour-off-diagonal contributions to the down Yukawa coupling. A first consequence is the renormalization of CKM elements involving one heavy quark [85]. Further, non-holomorphic flavour-changing couplings result in $\tan\beta$-enhanced down-quark FCNC [90, 91]. While these are strongly CKM-suppressed for $s-d$ transitions, they yield sizeable effects in $b-d$ and $b-s$ transitions. Phenomenologically, large-$\tan\beta$ FCNC are relevant in $B_{d,s} \to \mu^+\mu^-$ [92] and $\Delta M_{d,s}$ [93], as well as in $B \to X_s \gamma$ [94].

A first hint to the numerically allowed range of $\tan\beta$ is given by the requirement of perturbative Yukawa couplings above the electroweak scale. The top Yukawa coupling $y_t \sim 1/\sin\beta$ sets a rough lower bound of $\tan\beta \gtrsim 1.5$ in the MSSM and of $\tan\beta \gtrsim 2.5$ in SO(10) GUTs, cf. Fig. 2.2. Similarly, one derives an upper bound of $\tan\beta \lesssim 65$ from the bottom Yukawa coupling $y_b \sim 1/\cos\beta$, depending on the size of the $\tan\beta$-enhanced corrections in Eq. (4.6). To gain further information on the Higgs sector of the MSSM, it is instructive to confront the various $\tan\beta$-enhanced effects with experiment. This leads to constraints in the two-parameter space $(\tan\beta, M_{H^+})$, which determines the Higgs mass spectrum at tree level, cf. Eq. (1.25). The usual quest is to pin down the mass of the

---
[1] Peccei and Quinn originally studied a chiral U(1) invariance within the context of CP conservation in strong interactions.

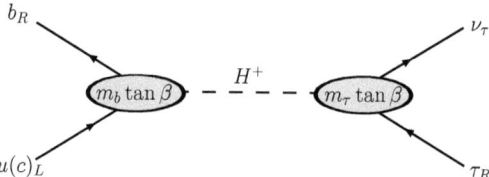

Figure 4.2: Doubly $\tan\beta$-enhanced charged-Higgs countributions to $B \to (D)\tau\nu$.

charged Higgs boson for a fixed value of $\tan\beta$. To this end, two complementary roads are pursued: Direct Higgs boson searches at colliders and indirect analyses of flavour physics observables. Direct constraints on $(\tan\beta, M_{H^+})$ arise from the combination of neutral-Higgs searches (mostly in the process of Higgsstrahlung) and charged-Higgs searches (in $t \to bH^+$). The reach of the experiments at LEP and TeVatron excludes light charged Higgs bosons with $M_{H^+} \lesssim 150$ GeV, largely independently from $\tan\beta$ [95]. Indirect constraints from B physics observables generally cut stronger into the $(\tan\beta, M_{H^+})$ parameter space. Recent analyses exploit the correlations between large-$\tan\beta$ effects in B physics and combine them with flavour-independent direct constraints [96–98]. Since these studies largely rely on loop-induced FCNC, they involve a large number of SUSY parameters. Therefore the resulting constraints on $(\tan\beta, M_{H^+})$ highly depend on the respective scenario. Processes involving three-level charged-Higgs exchange provide cleaner and widely model-independent constraints, since they give access to the quantity $\tan\beta/M_{H^+}$ with reduced sensitivity to other MSSM parameters. In the following, we will concentrate on the decay modes $B \to \tau\nu$ and $B \to D\tau\nu$ to derive rigorous constraints on the coupling of charged Higgs bosons to fermions.

4.2 Charged Higgs in $B \to (D)\tau\nu$ branching ratios

Within the last two years, the collaborations at the B factories BaBar and BELLE have made important progress in the analysis of the accumulated data to probe extensions of the Higgs sector of the Standard Model. In particular, the branching fractions for tauonic and semi-tauonic B decays, $\mathcal{B}(B \to \tau\nu)$ and $\mathcal{B}(B \to D\tau\nu)$, have been extracted from the data. $B \to \tau\nu$ and $B \to D\tau\nu$ share the feature of sensitivity to charged-Higgs currents. Since the couplings of right-handed b quarks and τ leptons to charged Higgs bosons are enhanced by $\tan\beta$, depicted in Fig. 4.2, both decay rates possibly receive significant contributions due to charged-Higgs exchange. In the scenario of large $\tan\beta$, the discovery potential for a relatively light charged Higgs boson in $B \to \tau\nu$ and $B \to D\tau\nu$ is high. Vice versa, non-observation sets strong constraints on M_{H^+} for fixed $\tan\beta$. Still, the two decay channels are not congruent in the search for charged Higgs bosons, due to both theoretical and experimental aspects. $B \to D\tau\nu$ compares to $B \to \tau\nu$ as follows:

4.2. Charged Higgs in $B \to (D)\tau\nu$ branching ratios

- The branching fraction $\mathcal{B}(B \to D\tau\nu)$ exceeds $\mathcal{B}(B \to \tau\nu)$ by about a factor of 50 in the Standard Model.

- $B \to D\tau\nu$ involves the well-known CKM element $|V_{cb}|$, whose uncertainty is much smaller than for the element $|V_{ub}|$, which governs $B \to \tau\nu$.

- Hadronic effects in the decay $B \to \tau\nu$ are parametrized by the B meson decay constant f_B, which must be obtained from non-perturbative methods. Current lattice gauge theory computations can determine f_B^2 with an uncertainty of roughly 20% [69]. $B \to D\tau\nu$ involves two hadronic form factors, one of which has been measured in $B \to D\ell\nu$ ($\ell = e, \mu$) decays [99, 100]. The other one is tightly constrained by Heavy Quark Effective Theory (HQET), such that hadronic effects are under control up to an uncertainty of less than 10%.

- Unlike $B \to \tau\nu$, the three-body decay $B \to D\tau\nu$ permits the study of decay distributions, which discriminate between W^+ and H^+ contributions. The novel prospects of differential distributions in $B \to D\tau\nu$ will be explained in Sec. 4.3.

- The SM contribution to the two-body decay $B \to \tau\nu$ is (mildly) helicity-suppressed, which enhances its sensitivity to charged-Higgs currents. A similar effect occurs in $B \to D\tau\nu$ near the kinematic endpoint, where the pseudoscalar D meson moves slowly in the B rest frame [101]: While the transverse modes $W_\perp^+$ of W^+ bosons suffer from a P-wave suppression, the virtual scalar H^+ recoils against the D meson in an unsuppressed S wave. Therefore $B \to D\tau\nu$ is preferable to $B \to D^*\tau\nu$, where the vector meson D^* in the final state prevents the suppression of transversal $W_\perp^+$ modes.

In order to quantify charged-Higgs effects in $B \to (D)\tau\nu$, let us elaborate on the theoretical aspects of the decay rates. The effective Hamiltonian describing $b \to q\tau\nu$, $q = u(c)$ transitions mediated by W^+ and H^+ exchange is given by

$$\mathcal{H}_{\text{eff}} = \frac{G_F}{\sqrt{2}} V_{qb} \left\{ [\bar{q}\gamma^\mu(1-\gamma_5)b] [\bar{\tau}\gamma_\mu(1-\gamma_5)\nu_\tau] \right. \\ \left. - \frac{\overline{m}_b m_\tau}{m_B^2} [\bar{q}(g_S + g_P\gamma_5)b] [\bar{\tau}(1-\gamma_5)\nu_\tau] \right\} + \text{h.c.} \tag{4.7}$$

The above operators, as well as $\overline{m}_b$, are defined in the $\overline{\text{MS}}$ scheme. Note the relative sign between the W^+ and H^+ contributions, which renders the interference destructive. This leads to a suppression of the branching ratios $\mathcal{B}(B \to (D)\tau\nu)$ with respect to the SM, as long as the charged-Higgs contribution does not dominate the decay rate. The meson mass m_B is factored out in the charged-Higgs coupling, so that $\mathcal{B}(B \to \tau\nu)$ vanishes for $g_P = 1$, see Eq. (4.10) below. The pseudoscalar coupling constant g_P governs charged-Higgs effects in $B \to \tau\nu$, while the scalar coupling g_S enters $B \to D\tau\nu$ only. In the MSSM one has $g_P = g_S$. By introducing these effective couplings, it is possible to perform a scenario-independent study of charged-Higgs couplings. In the two-Higgs-doublet model of type II,

$f_B =$	$(200 \pm 20)\,\text{MeV}$	[69]	$\tau_B =$	$(1.638 \pm 0.011)\,\text{ps}$	[54]
$f_{\pi^-} =$	$(130.4 \pm 0.04 \pm 0.2)\,\text{MeV}$	[54]	$\tau_\tau =$	$(290.6 \pm 1.0)\,\text{fs}$	[54]
$|V_{ub}| =$	$(3.87 \pm 0.09 \pm 0.46) \cdot 10^{-3}$	[104]	$\overline{m}_c/\overline{m}_b =$	0.222 ± 0.004	[73, 105]
$|V_{cb}| =$	$(41.6 \pm 0.6) \cdot 10^{-3}$	[54]	$\alpha_s(M_Z) =$	0.1189 ± 0.001	[106]
$|V_{ud}| =$	0.97418 ± 0.00027	[54]			

Table 4.1: Input values for $B \to \tau\nu$ and $B \to D\tau\nu$ branching ratios and decay distributions.

$g_{P,S} = m_B^2 \tan^2\beta / M_{H^+}^2$. In the framework of the MSSM, however, supersymmetric Yukawa corrections modify the charged-Higgs couplings to be

$$g_S = g_P = \frac{m_B^2}{M_{H^+}^2} \frac{\tan^2\beta}{(1 + \epsilon_{\tilde{g}} \tan\beta)(1 + \epsilon_\tau \tan\beta)}. \tag{4.8}$$

This expression holds in a scenario with minimal flavour violation. The corrections ϵ_τ to the τ lepton Yukawa coupling stem from neutralino-slepton and chargino-sneutrino loops given in Ref. [102], which are numerically smaller than the gluino contributions $\epsilon_{\tilde{g}}$. Since the Yukawa corrections are enhanced by a factor of $\tan\beta$, they can significantly change the constraints on $(\tan\beta, M_{H^+})$ with respect to the 2HDM of type II. In particular, $\epsilon_{\tilde{g}}$ and ϵ_τ can receive a complex phase from the μ parameter if first-generation sfermions are sufficiently heavy to soften the bounds from electric dipole moments on $\arg\mu$. Beyond minimal flavour violation, the phases in the sfermion mass matrices render Yukawa corrections complex. It is therefore mandatory to measure both the magnitude and phase of g_S.

$B \to \tau\nu$ branching fraction

The branching ratio of $B \to \tau\nu$ is calculated from the effective Hamiltonian in Eq. (4.7) by taking its matrix element squared and subsequently integrating over the two-body phase space of the final state. The matrix elements of the relevant axial-vector and pseudo-scalar quark currents are parametrized in terms of the B meson decay constant f_B,

$$\langle 0 | \bar{u}\gamma^\mu\gamma_5 b | B^- \rangle = -i f_B p_B^\mu, \qquad \langle 0 | \bar{u}\gamma_5 b | B^- \rangle = i f_B \frac{m_B^2}{\overline{m}_b}, \tag{4.9}$$

where p_B is the four-momentum carried by the B^- meson. The branching ratio finally reads [103]

$$\mathcal{B}(B \to \tau\nu) = \frac{G_F^2}{8\pi} \tau_B |V_{ub}|^2 f_B^2 m_B m_\tau^2 \left(1 - \frac{m_\tau^2}{m_B^2}\right)^2 (1 - g_P)^2, \tag{4.10}$$

with the lifetime of the B meson, τ_B. Using the input values from Tab. 4.1, our estimate for $\mathcal{B}(B \to \tau\nu)$ in the Standard Model is [2]

$$\mathcal{B}(B \to \tau\nu)^{\text{SM}} = (1.06^{+0.56}_{-0.40}) \cdot 10^{-4}, \tag{4.11}$$

---

[2] The uncertainty is calculated using flat errors on the input parameters.

4.2. Charged Higgs in $B \to (D)\tau\nu$ branching ratios

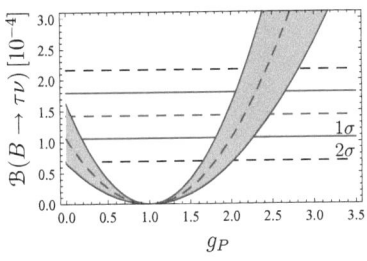 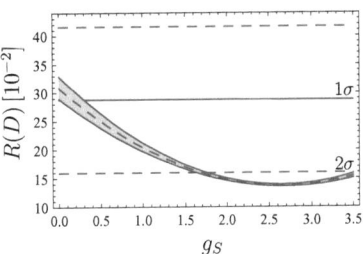

Figure 4.3: Branching fractions $\mathcal{B}(B \to \tau\nu)$ (left) and $R(D) = \mathcal{B}(B \to D\tau\nu)/\mathcal{B}(B \to D\ell\nu)$ (right) as a function of g_P and g_S, respectively. Light gray: Experimental one-sigma range. Dark gray: Theory estimations based on Eqs. (4.10) and (4.16).

which is in agreement with the average by the Heavy Flavour Averaging Group (HFAG) of the measurements at BaBar and BELLE [63],

$$\mathcal{B}(B \to \tau\nu)^{\text{exp}} = (1.43 \pm 0.37) \cdot 10^{-4}. \tag{4.12}$$

In Fig. 4.3 left, we confront $\mathcal{B}(B \to \tau\nu)^{\text{exp}}$ with the theory estimation as a function of the charged-Higgs coupling g_P. The measurement provides the following constraints on g_P,

$$g_P < 0.34 \;\cup\; 1.66 < g_P < 2.81 \qquad 95\% \text{ CL.} \tag{4.13}$$

$B \to D\tau\nu$ branching fraction

The calculation of the branching ratio for $B \to D\tau\nu$ requires the matrix elements of the vector and scalar quark currents between the B and D meson bound states, which are parametrized by two form factors F_V and F_S,

$$\langle D(p_D)| \bar{c}\gamma^\mu b |\bar{B}(p_B)\rangle = F_V(q^2)\left[p_B^\mu + p_D^\mu - m_B^2 \frac{1-r^2}{q^2} q^\mu\right] + F_S(q^2) m_B^2 \frac{1-r^2}{q^2} q^\mu,$$
$$\langle D(p_D)| \bar{c} b |\bar{B}(p_B)\rangle = \frac{m_B^2(1-r^2)}{\overline{m}_b - \overline{m}_c} F_S(q^2). \tag{4.14}$$

Here, p_B and p_D denote the meson four-momenta, $q = p_B - p_D$ is the momentum transfer to the leptons, and $r = m_D/m_B$ is the ratio of D and B meson masses. The vector form factor $F_V(q^2)$ parametrizes contributions to the matrix element from transverse $W_\perp^+$ modes, whereas the scalar form factor $F_S(q^2)$ comes with longitudinal $W_\parallel^+$ modes and H^+ contributions. It is convenient to describe the kinematics of the decay by the dimensionless variable

$$w \equiv (1 + r^2 - q^2/m_B^2)/2r, \qquad 1 \leq w \leq w_{\text{max}} = \frac{m_B^2 + m_D^2 - m_\ell^2}{2 m_B m_D}, \tag{4.15}$$

where m_ℓ is the mass of the lepton in the final state of a semileptonic $B \to D$ decay. The kinematic endpoint of maximal momentum transfer q^2, where the D meson is produced at rest, is associated with $w = 1$. In the rest frame of the B meson, $w = E_D/m_D$ is the normalized kinetic energy of the D meson. One obtains the differential decay rate for $B \to D\tau\nu$ in terms of the variable w [107],

$$\frac{d\Gamma(B \to D\tau\nu)}{dw} = \frac{d\Gamma_\ell}{dw}\left(1 - \frac{r_\tau^2}{t}\right)^2 \left\{1 + \frac{r_\tau^2}{2t} + \frac{3r_\tau^2(1-r^2)^2}{8tr^2(w^2-1)}\left(1 - \frac{g_S t(w)}{1 - \overline{m}_c/\overline{m}_b}\right)^2 \frac{F_S(w)^2}{F_V(w)^2}\right\},$$

$$\frac{d\Gamma_\ell}{dw} = \frac{G_F^2}{12\pi^3}|V_{cb}|^2 m_B^5\, r^4 (w^2-1)^{3/2} F_V(w)^2\,, \qquad (4.16)$$

where $d\Gamma_\ell/dw$ is the differential decay rate for $B \to D\ell\nu$ with $m_\ell = 0$, $r_\tau = m_\tau/m_B$, and $t(w) = 1 + r^2 - 2rw = q^2/m_B^2$. Details on the calculation can be found in Ref. [108]. The branching ratio is then derived by integrating the differential rate within the kinematical limits given in Eq. (4.15). The experiments at the B factories extract $\mathcal{B}(B \to D\tau\nu)$ normalized to the decay into light leptons [109],

$$R(D)^{\mathrm{exp}} \equiv \frac{\mathcal{B}(B \to D\tau\nu)}{\mathcal{B}(B \to D\ell\nu)} = (41.6 \pm 11.7 \pm 5.2) \cdot 10^{-2}. \qquad (4.17)$$

The normalization to $\mathcal{B}(B \to D\ell\nu)$ reduces the dependence on $|V_{cb}|F_V$ and thereby the overall uncertainty. The uncertainty of about 30% is comparable with the error for the measurement of $\mathcal{B}(B \to \tau\nu)$. With the input values given in Tab. 4.1, we obtain the SM value within the experimental one-sigma range,

$$R(D)^{\mathrm{SM}} = (30.8^{+2.0}_{-1.9}) \cdot 10^{-2}. \qquad (4.18)$$

We use $|V_{cb}|$ determined from inclusive semileptonic B decays. For the parametrization and the input values of the form factors F_V and F_S entering $R(D)$ we refer to Sec. 4.3, where we will perform a careful analysis.[3] In Fig. 4.3 right, we plot $R(D)$ as a function of g_S in comparison with the experimental result. Despite the good control of uncertainties on the theory side, $R(D)$ does not allow strong constraints on g_S, because $\mathcal{B}(B \to D\tau\nu)$ is less sensitive to charged-Higgs contributions than $\mathcal{B}(B \to \tau\nu)$. $R(D)$ provides the constraints

$$g_S < 1.66\ \cup\ 3.63 < g_S \qquad 95\%\ \mathrm{CL}. \qquad (4.19)$$

For $g_P = g_S$, one derives the combined constraints on the charged-Higgs coupling from $\mathcal{B}(B \to \tau\nu)$ in Eq. (4.13) and $R(D)$,

$$g_S < 0.34 \qquad 95\%\ \mathrm{CL}. \qquad (4.20)$$

Note that $R(D)$ excludes the range around $g_S = 2$ left by $\mathcal{B}(B \to \tau\nu)$. The constraints on charged-Higgs contributions in $\mathcal{B}(B \to \tau\nu)$ and $R(D)$ are visualized in the $(\tan\beta, M_{H^+})$ plane in Fig. 4.4. We display the excluded parameter regions in the 2HDM (left) and

---
[3] The form factor inputs are given in Eqs. (4.26) and (4.30).

4.2. Charged Higgs in $B \to (D)\tau\nu$ branching ratios

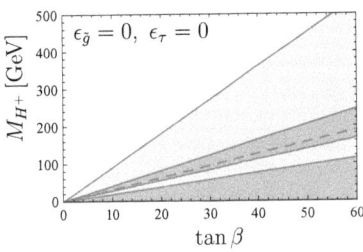
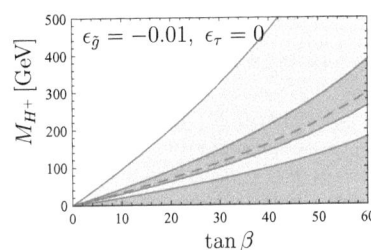

Figure 4.4: Excluded regions at 95% CL in the $(\tan\beta, M_{H^+})$ plane from $\mathcal{B}(B \to \tau\nu)$ (light gray) and $R(D)$ (dark gray) in the 2HDM (left) and in the MSSM (right). The dashed lines show the contours of the areas excluded by $\mathcal{B}(B \to \tau\nu)$.

the MSSM (right), using the charged-Higgs coupling from Eq. (4.8) for a typical amount of Yukawa corrections. The corrections ϵ_τ to the $H^+ - \tau$ coupling are neglected. Via the constraints on $(\tan\beta, M_{H^+})$, $B \to (D)\tau\nu$ decays provide important information for the direct discovery of a charged Higgs boson at colliders: In a scenario with large $\tan\beta$, the charged Higgs must be quite heavy, especially if supersymmetric Yukawa corrections enhance its coupling to fermions. The discovery of a H^+ in the $B \to (D)\tau\nu$ branching fractions themselves is possible, but difficult. From Fig. 4.3 one sees that a high experimental precision is needed in order to distinguish charged-Higgs contributions in the decay rates. In particular, the extraction of g_S is complicated in $\mathcal{B}(B \to \tau\nu)$ by the considerable theoretical uncertainty, and in $\mathcal{B}(B \to D\tau\nu)$ by the flat dependence on the charged-Higgs coupling.

Finally, a comment on charged-Higgs effects in other leptonic meson decay modes like $D_s \to \tau\nu$ and $K \to \mu\nu$ is in order. Since the branching fractions and decay constant for these decay modes are known to much better precision than for $B \to \tau\nu$, it is worth to compare the respective H^+ – fermion couplings. In the 2HDM, the charged-Higgs coupling in the leptonic decay of a meson $M = \bar{d}_j u_i$ is given by

$$g_P^M = \frac{m_M^2}{M_{H^+}^2} \frac{m_{d_j}}{m_{u_i} + m_{d_j}} \tan^2\beta, \tag{4.21}$$

independently of the flavour of the lepton in the final state. This translates into the explicit couplings in D_s and K decays,

$$g_P^{D_s} = \frac{m_{D_s}^2}{M_{H^+}^2} \frac{m_s}{m_c + m_s} \tan^2\beta \approx 0.002\, g_P^B, \qquad g_P^K = \frac{m_K^2}{M_{H^+}^2} \tan^2\beta \approx 0.009\, g_P^B. \tag{4.22}$$

Due to this significant suppression with respect to B decays, only $K \to \mu\nu$ reaches the level of $B \to D\tau\nu$ concerning constraints in the $(\tan\beta, M_{H^+})$ plane in Fig. 4.4. Thus, none of the leptonic decays of lighter mesons provides additional information on $(\tan\beta, M_{H^+})$ to the constraints from the branching ratios of $B \to \tau\nu$ and $B \to D\tau\nu$.

4.3 $B \to D\tau\nu$ differential distributions

Even though the branching fraction of $B \to D\tau\nu$ is less sensitive to charged-Higgs effects than $\mathcal{B}(B \to \tau\nu)$, the semi-tauonic mode is valuable, since it allows to study differential decay distributions. These can distinguish H^+ from W^+ contributions through the shape of the decay spectrum [101,107,110]. Thereby it is possible to not only constrain, but to measure $|g_S|$ and to additionally determine a possible CP phase from Yukawa corrections in the charged-Higgs coupling. A first step in this direction involves the D meson energy spectrum $d\Gamma(B \to D\tau\nu)/dw$ from Eq. (4.16), which probes the different dependence on the momentum transfer q^2 of H^+ and W^+ contributions. To gain further information about the characteristics of charged-Higgs couplings, one has to investigate the polarization of the τ lepton. Since the taus emerging from a H^+ decay are dominantly right-handed, the angular distribution of the decay products in the final state will change characteristics in presence of charged-Higgs contributions [111]. However, the reconstruction of the τ lepton is challenging at the B factories, because it decays too fast for a displaced vertex and its decay involves at least one more neutrino. In particular, the τ polarization is not directly accessible. The straightforward way to deal with the missing information on the τ kinematics is to study the full decay chain down to the final detectable particles stemming from the τ lepton. The resulting decay distribution preserves the dependence on the τ polarization and thereby exhibits an increased sensitivity to g_S with respect to $d\Gamma/dw$. We have studied the decays $\tau^- \to \ell^- \bar{\nu}_\ell \nu_\tau$, $\tau^- \to \pi^- \nu_\tau$, and $\tau^- \to \rho^- \nu_\tau$ and find the decay chain $\bar{B} \to D\bar{\nu}_\tau \tau^- [\to \pi^- \nu_\tau]$ most powerful to discriminate between H^+ and W^+ contributions. This can be understood by comparing the energy spectrum of the particles ℓ^-, π^-, and ρ^- in the final state, displayed in Fig. 1 of Ref. [111]. Energetic pions stem dominantly from τ_R decays (H^+ and longitudinal $W_\parallel^+$), while soft pions originate from τ_L (transversal $W_\perp^+$). For muons and ρ mesons the spectra are less characteristic. We therefore will elaborate on the experimentally accessible triple differential distribution

$$d^3\Gamma \equiv \frac{d\Gamma(B \to D\nu\tau[\to \pi\nu])}{dE_D \, dE_\pi \, d\cos\theta_{D\pi}}, \quad (4.23)$$

which retains information on the τ polarization through the explicit dependence on the π^- energy E_π and the angle $\theta_{D\pi}$ between the meson momenta $\vec{p}_D$ and $\vec{p}_\pi$. We define these quantities in the rest frame of the B meson.

Confronting the above considerations with the experimental feasibility, one learns that the price to pay for an increased sensitivity to charged-Higgs effects is a significant loss of statistics. The current amount of data collected at the B factories allows the projection on the q^2 distribution, cf. Ref. [109], so that a fit of $d\Gamma(B \to D\tau\nu)/dw$ to g_S comes into reach. The triple distribution $d^3\Gamma$ is more difficult to handle, because the fraction of τ decays into $\pi^-\nu_\tau$ amounts to about 10% only. Still, this sensitive observable provides a promising tool to measure the charged-Higgs coupling with increased statistics from upgraded B experiments. In the following, we thus discuss the prospects of both the E_D spectrum $d\Gamma/dw$ and the triple differential rate $d^3\Gamma$, which mark the way to quantify g_S from $B \to D\tau\nu$ distributions.

Hadronic form factors for $B \to D$ transitions

In order to maximize the theoretical precision of the differential decay distributions in $B \to D\tau\nu$, it is crucial to know both the normalization and the shape of the two form factors $F_V(q^2)$ and $F_S(q^2)$ given in Eq. (4.14). To avoid a pole in the decay amplitude at $q^2 = 0$, the form factors have to be equal at maximal D meson recoil,

$$F_V(q^2 = 0) = F_S(q^2 = 0). \tag{4.24}$$

At the opposite point of the spectrum, namely at $w = 1$, the vector and scalar form factors are related within the framework of Heavy Quark Effective Theory (HQET) [112–115]. In HQET, flavour and spin symmetries between mesons that contain one heavy quark Q are manifest in the limit $m_Q \to \infty$. In particular, in the limit of heavy-quark symmetry all form factors in $B \to D^{(*)}$ transitions reduce to the universal Isgur-Wise function $\xi(w)$ [116]. At the leptonic endpoint, this function is normalized to $\xi(w = 1) = 1$. Deviations from the heavy-quark limit can be expressed by a series of corrections of $\mathcal{O}(1/m_Q, \alpha_s)$. The HQET description of form factors is justified if the momentum transfer to the light quark in the meson is small compared to the heavy quark mass m_Q. This condition is fulfilled in the kinematically allowed range of $B \to D\tau\nu$ decays from Eq. (4.15). To apply heavy-quark constraints to $B \to D$ transitions, we define the rescaled vector and scalar form factors $V_1(w) \equiv F_V(w) \cdot 2\sqrt{r}/(1+r)$ and $S_1(w) \equiv F_S(w) \cdot (1+r)/\sqrt{r}\,(w+1)$. At $w = 1$, the heavy-quark relation reads

$$V_1(1) = S_1(1) = 1 + \mathcal{O}(1/m_{c,b}, \alpha_s). \tag{4.25}$$

Let us focus on the scalar form factor S_1. At the leptonic endpoint, $S_1(w = 1)$ is protected from $\mathcal{O}(1/m_Q)$ corrections to the heavy-quark limit, according to Luke's theorem [117]. The radiative corrections to heavy-quark currents of $\mathcal{O}(\alpha_s)$ have been calculated in Ref. [118]. We add an uncertainty of 5% to account for corrections of $\mathcal{O}(1/m_{c,b}^2, \alpha_s^2)$ and get[4]

$$S_1(1) = 1.02 \pm 0.05. \tag{4.26}$$

The shape of the form factors can be described in terms of only two parameters by exploring dispersion relations and analyticity properties. There are different parametrizations on the market [119, 120], which however are based on the same techniques and therefore lead to very similar results [2]. After a conformal mapping of the kinematic variable

$$\frac{q^2}{m_B^2} = t \longrightarrow z(t, t_0) = \frac{\sqrt{t_+ - t} - \sqrt{t_+ - t_0}}{\sqrt{t_+ - t} + \sqrt{t_+ - t_0}} \tag{4.27}$$

with $|z| < 1$ and $t_\pm = (1 \pm r)^2$, the form factors can be expanded in a power series of z,

$$F_{V,S}(t) = \sum_{k=0}^{\infty} C_k^{V,S}(t_0)\, z^k(t, t_0). \tag{4.28}$$

---

[4]We compute $S_1(1) = \hat{C}_1 - \hat{C}_2 - \hat{C}_3$ using the Wilson coefficients $\hat{C}_i$ for $w = 1$ from the appendix of Ref. [118]. The quark mass inputs required for the determination of $\alpha_s(m_{c,b})$ are taken from Ref. [105].

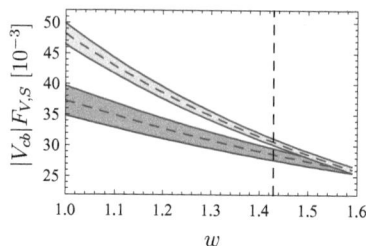

Figure 4.5: Form factors $|V_{cb}|F_V(w)$ (light gray, cf. Eq. (4.30)) and $|V_{cb}|F_S(w)$ (dark gray, cf. Eqs. (4.24), (4.26), and (A.26)) for semileptonic $B \to D$ decays. The vertical dashed line marks the endpoint of maximal recoil in $B \to D\tau\nu$, $w^{\max} = 1.43$.

Choosing $t_0 = t_-$, the vector form factor is parametrized to very good approximation by [119]

$$V_1(t) = \mathcal{G}(1) \cdot \left\{ 1 - 8\rho^2 \, z(t,t_0) + (51\rho^2 - 10) \, z^2(t,t_0) - (252\rho^2 - 84) \, z^3(t,t_0) \right\}. \quad (4.29)$$

From the differential distribution in Eq. (4.16), one learns that the decay $B \to D\ell\nu$ involves only the vector form factor F_V if the lepton in the final state is massless. The collaborations BaBar and BELLE have recently extracted the normalization $|V_{cb}|\mathcal{G}(1)$ and the shape parameter ρ^2 of the vector form factor from the spectrum $d\Gamma_\ell/dw$ [99, 100]. The world average by the HFAG reads [63]

$$|V_{cb}|\mathcal{G}(1) = (42.3 \pm 0.7 \pm 1.3) \cdot 10^{-3}, \qquad \rho^2 = 1.18 \pm 0.04 \pm 0.04, \quad (4.30)$$

with a correlation of 0.88. The scalar form factor is thereby automatically known at $t = 0$ from Eq. (4.24). Using the HQET value at the opposite edge of the spectrum from Eq. (4.26) as second input, F_S is fixed over the entire kinematic range by a two-parameter function based on Eq. (4.28). The explicit expression is given in Appendix A.4.

The form factors resulting from these considerations are shown in Fig. 4.5. Uncertainties are largest at $w = 1$. The vector form factor $|V_{cb}|F_V(w)$ is thereby known to a precision of less than $\delta|V_{cb}|F_V(w = 1) = 3.5\%$. The error of the scalar form factor is dominated by the 5% uncertainty on the HQET input and amounts to $\delta|V_{cb}|F_S(w = 1) = 6.4\%$. These results provide good prospects to distinguish charged-Higgs contributions in the differential distributions of $B \to D\tau\nu$.

An alternative route to pin down hadronic uncertainties in scalar currents might be given by lattice calculations. In the quenched approximation, the ratio $F_S(w)/F_V(w)$, parametrized by $\Delta(w)$, has been computed within the kinematic range $1 < w < 1.2$ with an error of 2% [121, 122]. The $B \to D\tau\nu$ decay distributions depend only on this ratio, once one normalizes to the $B \to D\ell\nu$ spectrum, cf. Eq. (4.16). By using the lattice input

4.3. $B \to D\tau\nu$ differential distributions

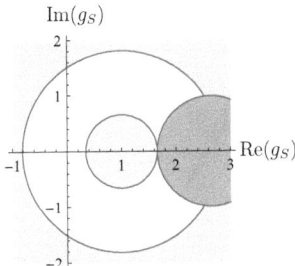

Figure 4.6: Exclusions in the complex g_S plane from $\mathcal{B}(B \to \tau\nu)$ (light gray) and $R(D)$ (dark gray) at 95% CL, assuming $g_S = g_P$.

$\Delta(1) = 0.46 \pm 0.01$ instead of the HQET value for $S_1(1)$, one could improve the overall precision of scalar contributions in the normalized $B \to D\tau\nu$ spectrum. However, further uncertainties are expected due to the extrapolation of $\Delta(w)$ to larger values of w and the suppression of sea-quark masses in the simulation. Therefore we do not find the lattice approach preferable to our conservative error estimation based on HQET.

D meson energy spectrum

The branching ratios of $B \to \tau\nu$ and $B \to D\tau\nu$ decays considered in Sec. 4.2 put already stringent constraints on the absolute value of the charged-Higgs coupling g_S. They are, however, not sensitive to a potential phase of g_S. We illustrate the constraints on g_S from $\mathcal{B}(B \to \tau\nu)$ and $R(D)$ in the complex plane in Fig. 4.6. The real axis reflects the results from Sec. 4.2, see Eqs. (4.13) and (4.19). The constraint from the $B \to D\tau\nu$ branching ratio removes part of the white ring left by $B \to \tau\nu$.
Further constraints can be added by differential distributions of $B \to D\tau\nu$, since they allow to discriminate between different values of g_S through the shape of the spectrum. Let us once more come back to the D meson energy spectrum in Eq. (4.16). Close to the leptonic endpoint, the transverse modes $W_\perp^+$ (F_V) are suppressed with respect to longitudinal $W_\parallel^+$ and charged-Higgs contributions (F_S) by $(w^2 - 1)$. This is the analytic explanation for the P-wave suppression of transverse modes mentioned in the introduction to this section. Further, charged-Higgs contributions exhibit an additional dependence on w, which distinguishes them from $W_\parallel^+$ modes. Experimentally, the latter feature is more important than the $W_\perp^+$ suppression near $w = 1$, because of the poor statistics at the leptonic endpoint. Via a fit to the entire spectrum, it is possible to gain valuable additional information on g_S. In Fig. 4.7 left, we show the spectrum $d\Gamma(B \to D\tau\nu)/dw$ in the Standard Model and with charged-Higgs contributions. The exemplary values $g_S = 0.3$ and $g_S = 1 + 0.7i$ are allowed, but indistinguishable in $B \to (D)\tau\nu$ branching fractions. They lead to the same $\mathcal{B}(B \to \tau\nu)$, but have clearly different shapes in the w spectrum.

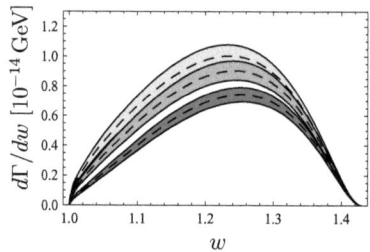 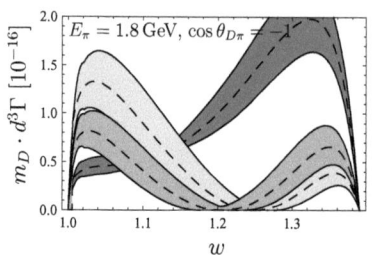

Figure 4.7: Differential decay distributions $d\Gamma(\overline{B} \to D^+\tau^-\bar{\nu}_\tau)/dw$ (left) and $d\Gamma(\overline{B} \to D^+\bar{\nu}_\tau\tau^-[\to \pi^-\nu_\tau])/dw\, dE_\pi d\cos\theta_{D\pi}$ (right) in the SM (light gray) and with charged-Higgs contributions. The displayed values $g_S = 0.3$ (gray) and $g_S = 1 + 0.7i$ (dark gray) are allowed by $\mathcal{B}(B \to \tau\nu)$ and $R(D)$ from Sec. 4.2.

One observes that it is difficult to strengthen the upper bound on g_S from the shape, because the sensitivity is not sufficient to detect tiny charged-Higgs effects with $g_S \lesssim 0.34$. However, in presence of a complex phase, charged-Higgs effects can change the shape of the spectrum significantly, as for $g_S = 1 + 0.7i$. Knowing the phase of g_S would give us insight into the structure of supersymmetric Yukawa corrections to the charged-Higgs coupling.

Triple differential distribution

To include the information on the τ polarization, we suggested to investigate the triple differential distribution $d^3\Gamma$ from Eq. (4.23) in the decay chain $\overline{B} \to D\bar{\nu}\tau[\to \pi^-\nu]$. The experimental access to this observable is given by the measurement of the energies E_D and E_π in the B rest frame, as well as the angle $\theta_{D\pi}$ between the mesons in the final state. The τ polarization is encoded in these quantities via the correlation with the momentum of the pion originating from the τ decay. If the τ lepton is right-handed, the pion preferably flies into the direction of the τ, as we illustrate in Fig. 4.8. Pions from left-handed τ leptons prefer the opposite direction. Charged Higgs bosons thereby manifest themselves in an excess of energetic pions, especially if the D and π mesons fly back-to-back. Background from hadronic decays à la $\overline{B} \to DD^-[\to \pi^-\pi^0]$ with unobserved neutral pions can be rejected by cuts in the angular distribution around

$$\cos\theta_{D\pi} = \frac{(m_B - E_D - E_\pi)^2 - 2(E_D^2 - m_D^2) - m^2}{2(E_D^2 - m_D^2)}, \quad (4.31)$$

where m is the mass of the undetected particle. We obtain the triple differential distribution by integrating over the neutrinos,

$$\frac{d\Gamma(\overline{B} \to D\bar{\nu}_\tau\tau^-[\to \pi^-\nu_\tau])}{dE_D\, dE_\pi\, d\cos\theta_{D\pi}} = G_F^4 f_\pi^2 |V_{ud}|^2 |V_{cb}|^2 \tau_\tau \quad (4.32)$$
$$\times \left[C_W(F_V, F_S) - C_{WH}(F_V, F_S)\, Re[g_S] + C_H(F_S)|g_S|^2 \right],$$

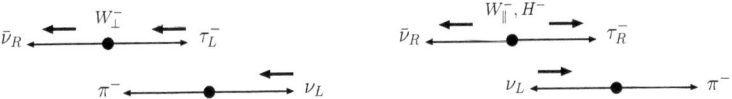

Figure 4.8: Correlation between the τ polarization and the π^- momentum in the decay chain $\overline{B} \to D\bar{\nu}\tau^-[\to \pi^-\nu]$.

with the functions C_W for the SM, C_{WH} for interference, and C_H for Higgs contributions. Explicit expressions are given in Appendix A.4. Inputs for the pion decay constant f_{π^-}, the CKM elements $|V_{ud}|$ and $|V_{cb}|$, and the τ lepton lifetime τ_τ are summarized in Tab. 4.1. The dependence of $d^3\Gamma$ on both $\text{Re}(g_S)$ and $|g_S|$ allows to distinguish the phase of the charged-Higgs coupling from the shape, as was already the case in the E_D spectrum $d\Gamma/dw$. Since the differential distributions are CP-conserving quantities, a phase can be extracted up to a twofold ambiguity. The increased sensitivity to charged-Higgs effects of $d^3\Gamma$ over $d\Gamma/dw$ becomes visible in Fig. (4.7) right. We plot $d^3\Gamma$ for fixed values $E_\pi = 1.8\,\text{GeV}$ and $\cos\theta_{D\pi} = -1$ as a function of w. This choice corresponds to a kinematical region with an energetic pion back-to-back with the D meson, where one expects H^+ contributions to show up. Compared to $d\Gamma/dw$, the additional information on the τ polarization changes the spectrum qualitatively. Thereby it is possible to measure the phase of g_S and furthermore to distinguish even tiny charged-Higgs effects through the shape of the spectrum. A maximum likelihood fit to $d^3\Gamma$ would automatically explore both the q^2 dependence and the τ polarization that characterize charged-Higgs contributions all over the phase space. Note that $B \to \tau\nu$ is not expected to resolve very small charged-Higgs effects. Even with significant improvements on the decay constant f_B and the CKM element $|V_{ub}|$, couplings of $g_P \simeq 0.2$ are very difficult to distinguish from the SM in the branching fraction. $B \to D\tau\nu$ distributions, if resolvable, definitively compete with $B \to \tau\nu$ and provide a complementary approach to H^+ searches at the LHC. Moreover, the distributions allow not only to find a charged Higgs boson, but to determine the magnitude and phase of its coupling to fermions.

4.4 Yukawa unification and $\tan\beta$

Knowing the value of $\tan\beta$ helps to gain insight into the structure of the Yukawa sector in an SO(10) framework. In particular, the unification of the third-generation Yukawa couplings at the SO(10) scale is very sensitive to $\tan\beta$. In the case of $\tan\beta \simeq 50$, y_t and y_b are of comparable size, allowing for top-bottom-tau unification at M_{10}. The RGE of the bottom Yukawa coupling, however, is affected by a priori large $\tan\beta$-enhanced threshold corrections near the electroweak scale, due to the supersymmetric loops from Fig. 4.1. To keep the Yukawa corrections $\epsilon_b^{\tilde{g}}$ and $\epsilon_b^{\tilde{\chi}}$ small, the relevant SUSY parameters ought to fulfill [84]

$$\mu\, m_{\tilde{g}},\ \mu\, A_t \ll m_{\tilde{q}}^2, \qquad (4.33)$$

leading to a spectrum with light fermionic and heavy scalar SUSY mass parameters. Importantly, bottom-tau unification itself strongly depends on $\tan\beta$, which enters the RGE of the bottom Yukawa coupling via y_t. To obtain a sufficiently low bottom mass, $\tan\beta$ has to be either of $\mathcal{O}(1)$ (which is excluded) or of $\mathcal{O}(50)$ [123]. Threshold corrections to m_b may relax this constraint to smaller values of $\tan\beta$, but would in turn spoil bottom-top unification.

In SO(10), top-bottom-tau unification is realized at M_{10} if $\tan\beta$ is large, and if the Yukawa couplings of all third-generation fermions are generated from a single term $Y^{ij}\, 16_i 16_j 10_H$ in the superpotential. Within this framework, fermion mixing and Yukawa corrections for light fermions are provided by adding either renormalizable terms with large Higgs representations 120_H and/or 126_H, or non-renormalizable higher-dimensional operators with $45_H \otimes 10'_H$, for instance. In the CMM model, top-bottom-tau unification is relaxed to potential bottom-tau unification, since up- and down-type Yukawa couplings are generated by separate terms in the superpotential. Due to the suppression of y_b with respect to y_t by v_{10}/M_{Pl}, which naturally implies $\tan\beta < 10$, bottom-tau unification works to about 18% only. The gauge couplings g'_1 and g_2 unify at $M_{\mathrm{GUT}} = 4 \cdot 10^{16}$ GeV, yielding $y_b(M_{\mathrm{GUT}}) = 0.83\, y_\tau(M_{\mathrm{GUT}})$ for the input value $\tan\beta = 5$ used in this work. This result is not very sensitive to the remaining CMM inputs $m_{\tilde{g}}$, $m_{\tilde{d}}$, a_d, and $\arg(\mu)$. More generally, bottom-tau unification is challenged in SO(10) models with small $\tan\beta$, where down-type Yukawa couplings are entirely generated by higher-dimensional operators. SU(5)-breaking terms can partially account for this mismatch via corrections of $v_5/v_{10} = \mathcal{O}(10\%)$. Another way out might be to relax the assumption $H_u \subset 10_H$, $H_d \subset 10'_H$ to linear combinations as in Eq. (2.5), such that the bottom Yukawa coupling receives a contribution from the large coupling to 10_H. In this case, however, the typical effects of large neutrino mixing among right-handed down squarks are more difficult to identify.

The potential unification of Yukawa couplings is a valuable and not at all obvious add-on to gauge coupling unification. Preserving Yukawa unification and simultaneously providing realistic fermion masses and mixings requires the refinement of the Yukawa sector by additional terms in any case. The perturbative SO(10) models built with higher-dimensional operators can be classified by means of $\tan\beta$. Large-$\tan\beta$ effects in flavour observables like in $B \to (D)\tau\nu$ or $B_s \to \mu^+\mu^-$ hint at top-bottom-tau unification at M_{10}. In absence of signals for large $\tan\beta$, atmospheric neutrino mixing effects in $B_{s,d}$ observables indicate the realization of CMM-like models with suppressed down-type Yukawa couplings. $B \to (D)\tau\nu$ decays play a special role to distinguish between different SO(10) models: Contrary to B_s observables, the constraints on $\tan\beta$ from $B \to (D)\tau\nu$ for a fixed SUSY spectrum are independent of atmospheric neutrino mixing effects. This allows us to figure out the realized model step by step from $b - s$ transitions and (semi)tauonic B decays.

Conclusions

In the attempt to answer grand questions by Grand Unification, flavour physics observables play a crucial role. They probe Yukawa unification independently from other hot spots in GUTs like proton decay. In this work, we focussed on the relation between right-handed down quarks and leptons. Their embedding in one single GUT representation translates lepton flavour mixing into flavour-changing neutral currents among right-handed down quarks. In particular, the large atmospheric neutrino mixing angle induces $b_R - s_R$ transitions. In the Standard Model, right-handed currents are generated only at loop level and strongly suppressed. In supersymmetric models, however, flavour mixing among the superpartners of right-handed down quarks can be large. Via flavour-changing down-quark-squark-gluino couplings, $\tilde{b}_R - \tilde{s}_R$ transitions translate effects of atmospheric neutrino mixing into B_s observables.

Perfect Yukawa unification can be realized only within the third generation of fermions. A GUT model with realistic fermion masses and mixings requires Yukawa corrections for light fermions. These are provided by adding higher-dimensional Yukawa terms suppressed by powers of $1/M_{\text{Pl}}$, with an a priori arbitrary flavour structure. Such terms may arise by integrating out heavy degrees of freedom at the Planck scale. Yukawa corrections introduce the large atmospheric neutrino mixing angle also into $b_R - d_R$ and $s_R - d_R$ transitions. This allows to probe the flavour structure of corrections to down-quark-lepton unification with K and B_d physics observables.

We have studied Yukawa (non)unification in the framework of the CMM model, a supersymmetric SO(10) model with flavour-blind SUSY breaking at the Planck scale. In this model, the Yukawa couplings of down quarks and leptons are generated by a dimension-five term and thereby suppressed with respect to up quarks by $v_{10}/M_{\text{Pl}} \sim 10^{-2}$, which naturally implies $\tan\beta \lesssim 10$. Imprints of Grand Unification in flavour physics observables are confined to effects of atmospheric neutrino mixing in down-squark currents. These effects are large in the case of a light gluino and a strong inverted mass hierarchy among down squarks, generated by the fast renormalization group evolution of Yukawa couplings in SO(10).

In Chapter 2, we have analyzed neutrino mixing effects in $B_s - \overline{B}_s$ mixing. The measured mass difference ΔM_s sets a lower limit on the gluino mass, $m_{\tilde{g}} \gtrsim 550\,\text{GeV}$, for a vanishing CMM phase $\phi_{B_s} = 0$, which limits the general magnitude of CMM effects. Still, the CP phase ϕ_s in $B_s - \overline{B}_s$ mixing can be significant. This finding can explain the discrepancy of $\gtrsim 2\sigma$ between the measurement of a sizeable ϕ_s and a very small phase in the Standard

Model. For $\phi_{B_s} \simeq 1$ rad and a small gluino mass, the resulting phase $\phi_s \simeq -0.5$ rad reaches the experimental one-sigma range.

The CMM model was extended in Chapter 3 to include Yukawa corrections to $Y_d = Y_e^\top$ of $v_5/v_{10} = \mathcal{O}(10^{-1})$. We parametrized the resulting additional rotations of d_R and s_R (s)quarks by a mixing angle θ and three phases. In the absence of accidental cancellations among the new phases, the CP-violating observable $|\epsilon_K|$ sets the strong constraint $\theta \lesssim 1°$. This implies that the flavour structure of Yukawa corrections to down-quark-lepton unification must be aligned with the initial unified couplings Y_d and Y_e. The mass relations $m_d - m_e$ and $m_s - m_\mu$ are therefore corrected without introducing new flavour-changing effects in B_d and other (less sensitive) K observables. Independently from kaon physics, the B physics observables ΔM_d, $S_{J/\psi K_S}$, and $\Delta M_s/\Delta M_d$ lead to the looser bound $\theta \lesssim 20°$, which applies for the case of a vanishing CMM phase in the kaon system, $\phi_K = 0$. As an application of neutrino mixing effects due to Yukawa corrections, we studied the determination of the unitarity triangle from K and B_d observables. The recently claimed tension concerning CP violation in the K and B_d systems can be removed within the CMM model. Taking into account the above-mentioned parameter constraints, CP-violating effects in K and B_d mixing are large enough to close the unitarity triangle and at the same time account for a sizeable CP phase in $B_s - \overline{B}_s$ mixing.

Beyond the CMM model, our results hold more generally in GUT models with small Higgs representations. Once the mass relations between light down-type quarks and leptons are corrected by higher-dimensional Yukawa terms, large neutrino mixing effects in $b_R - s_R$ transitions a priori imply effects in $b_R - d_R$ and $s_R - d_R$ transitions. In scenarios with minimal flavour violation, FCNC among down (s)quarks are made visible by breaking the universality of down-squark masses through renormalization group effects. While in the CMM model mass universality breaking stems from the large effects of y_t in SO(10) running, in SU(5) models off-diagonal mass matrix elements may arise from down-quark couplings to heavy Higgs bosons and right-handed neutrinos. Other sources of flavour and CP violation in $|\epsilon_K|$ would soften the constraints on θ. Yet, they would have to be fine-tuned to cancel the large neutrino mixing effects.

In Chapter 4, we addressed the impact of the parameter $\tan\beta$ on Yukawa unification in GUT models. Since successful top-bottom-tau unification requires $\tan\beta \simeq 50$, knowing its value is crucial to specify the Yukawa sector. The (semi)leptonic decays $B \to \tau\nu$ and $B \to D\tau\nu$ are sensitive to $\tan\beta$ in tree-level contributions of charged Higgs bosons. From the branching fractions of both decay modes the charged-Higgs coupling is constrained to $g_S \lesssim 0.34$ at 95% CL. The D meson energy spectrum $d\Gamma(B \to D\nu\tau)/dE_D$ can distinguish a potential phase in g_S if $\tan\beta/M_{H^+}$ is sufficiently large. Experimentally, a fit to the E_D spectrum is feasible with the present data collected at the B factories. With more statistics one could explore the triple differential decay rate $d\Gamma(B \to D\nu\tau[\to \pi\nu])/dE_D\, dE_\pi\, d\cos\theta_{D\pi}$. This observable contains additional information on the τ polarization, which distinguishes between the angular distributions of τ leptons stemming from W^+ and H^+ decays. The thereby increased sensitivity allows to measure the coupling of charged Higgs bosons for $g_S \lesssim 0.3$ through the shape of the spectrum, in addition to detecting a phase in the coupling. There are thus good prospects to discover charged Higgs bosons in the decay

Conclusions

distributions of $B \to D\tau\nu$ or, for a fixed SUSY spectrum, to pin down the value of $\tan\beta$. The information on $\tan\beta$ from $B \to (D)\tau\nu$ decays helps to reveal the Yukawa sector of a perturbative GUT model step by step. Knowing the magnitude of $\tan\beta$ clarifies whether down-type Yukawa couplings are generated from a higher-dimensional term (small $\tan\beta$) or unified with up-type couplings (large $\tan\beta$). Effects of large neutrino mixing in $b - s$, $b - d$, and $s - d$ transitions subsequently indicate the pattern of Yukawa corrections by constraining the parameters of a specific model.

Imprints of neutrino mixing in K and B physics observables probe the flavour structure of higher-dimensional Yukawa terms with different contributions to down quarks and leptons. While the effect of these terms on the mass relations $m_s - m_\mu$ and $m_d - m_e$ and on proton decay has been investigated before, the constraints on their flavour structure are a novel result of this work. The second class of Yukawa corrections, which contributes equally to down quarks and leptons, is relevant for proton decay. Our analysis is thus complementary to studies of higher-dimensional operators in proton decay. The finding of aligned Yukawa couplings and corrections supports the suppression of dimension-five contributions to proton decay in supersymmetric GUTs. Moreover, the flavour structure of the Yukawa sector is closely linked to the minimal Higgs field content of a model. Both the Yukawa and Higgs sectors therefore have to complement each other within a consistent and complete GUT model of flavour.

Appendix A

A.1 Weyl spinors

A four-component Dirac spinor ψ can be written in terms of two-component Weyl spinors ξ and χ,

$$\psi = \begin{pmatrix} \xi_\alpha \\ \epsilon^{\rho\sigma}\chi_\sigma \end{pmatrix}, \quad \bar\psi = (\chi^*_\beta \epsilon^{\beta\alpha}, \xi^*_\rho) \quad \text{with} \quad \epsilon^{\alpha\beta} = -\epsilon_{\alpha\beta} = 1, \quad \xi_\alpha = \epsilon_{\alpha\beta}\xi^\beta. \tag{A.1}$$

Lorentz indices α, β $(\rho, \sigma) = 1, 2$ are used for the upper (lower) components of ψ (and are not to be confounded with SU(2) indices). $\bar\psi = \psi^\dagger \gamma^0$ denotes the Dirac-conjugated spinor. We use the Weyl representation of the Dirac algebra,

$$\gamma^\mu = \begin{pmatrix} 0 & \sigma^\mu \\ \bar\sigma^\mu & 0 \end{pmatrix}, \quad \gamma_5 = \begin{pmatrix} -1 & 0 \\ 0 & 1 \end{pmatrix}, \tag{A.2}$$

with $\sigma^\mu, \bar\sigma^\mu$ given in Eq. (1.12). By applying the chirality projectors $P_{L,R} = (1 \mp \gamma_5)/2$ to the Dirac spinor,

$$P_L\psi = \begin{pmatrix} \xi_\alpha \\ 0 \end{pmatrix}, \quad P_R\psi = \begin{pmatrix} 0 \\ \epsilon^{\rho\sigma}\chi_\sigma \end{pmatrix}, \tag{A.3}$$

one learns that ξ is a "left-handed" and χ is a "right-handed" Weyl spinor. A charge-conjugated left-handed Dirac spinor reads

$$(\psi_L)^c = i\gamma^2 P_L \psi^* = i \begin{pmatrix} 0 & \sigma^2 \\ -\sigma^2 & 0 \end{pmatrix} \begin{pmatrix} \xi^*_\alpha \\ 0 \end{pmatrix} = -i \begin{pmatrix} 0 \\ \sigma^2 \xi^*_\rho \end{pmatrix}. \tag{A.4}$$

This is a right-handed spinor, and one verifies that $(\psi^c)_R = P_R i\gamma^2 \psi^* = (\psi_L)^c$. We thus define

$$\xi^c \equiv -i\sigma^2 \xi^*_\rho = \epsilon_{\rho\sigma}(\xi^*)^\sigma = -\epsilon^{\rho\sigma}(\xi^*)_\sigma. \tag{A.5}$$

Expressed in terms of Weyl spinors, the Lorentz scalar and vector bilinears read

$$\bar{\psi}_1 \psi_2 = (\chi_1^*)_\beta \, \epsilon^{\beta\alpha} (\xi_2)_\alpha + (\xi_1^*)_\rho \, \epsilon^{\rho\sigma} (\chi_2)_\sigma \equiv \chi_1^c \, \xi_2 + \xi_1^c \, \chi_2 \,,$$

$$\bar{\psi}_1 \gamma^\mu P_L \psi_2 = ((\chi_1^*)_\beta \, \epsilon^{\beta\alpha}, \; (\xi_1^*)_\rho) \begin{pmatrix} 0 & \sigma^\mu \\ \bar{\sigma}^\mu & 0 \end{pmatrix} \begin{pmatrix} (\xi_2)_\alpha \\ 0 \end{pmatrix} = \epsilon_{\rho\sigma} (\xi_1^*)^\sigma (\bar{\sigma}^\mu)^{\rho\alpha} (\xi_2)_\alpha \equiv \xi_1^c \, \bar{\sigma}^\mu \, \xi_2 \,,$$

$$\bar{\psi}_1 \gamma^\mu P_R \psi_2 = (\chi_1^*)^\beta \epsilon^{\beta\alpha} (\sigma^\mu)_{\alpha\rho} \epsilon^{\rho\sigma} (\chi_2)_\sigma = \chi_1^c \, \sigma^\mu \, \chi_2 \,.$$

(A.6)

One can identify ξ and χ with fermion fields f_L and f_R, such that for instance

$$\bar{\psi}_1 \psi_2 = (f_{1R})^c f_{2L} + (f_{1L})^c f_{2R} = (f_1^c)_L f_{2L} + (f_1^c)_R f_{2R} \,. \tag{A.7}$$

A.2 SO(10) decompositions

Irreducible decompositions The decomposition of SO(10) representations in terms of SU(5) fields reads

$$\begin{aligned}
16 &= 1 \oplus \bar{5} \oplus 10 \\
10 &= 5 \oplus \bar{5} \\
120 &= 5 \oplus \bar{5} \oplus 10 \oplus \overline{10} \oplus 45 \oplus \overline{45} \\
126 &= 1 \oplus \bar{5} \oplus 10 \oplus \overline{15} \oplus 45 \oplus \overline{50} \\
45 &= 1 \oplus 10 \oplus \overline{10} \oplus 24 \,.
\end{aligned} \tag{A.8}$$

The relations between tensor products and direct sums of SO(10) representations relevant for Yukawa couplings are

$$\begin{aligned}
16 \otimes 16 &= 10_S \oplus 120_A \oplus 126_S \\
10 \otimes 45 &= 10 \oplus 120 \oplus 320 \\
10 \otimes 16 &= \overline{16} \oplus \overline{144} \\
45 \otimes 16 &= 16 \oplus 144 \oplus 560 \,.
\end{aligned} \tag{A.9}$$

In SU(5), the irreducible decompositions of fermion bilinears are given by

$$\begin{aligned}
\bar{5} \otimes 10 &= 5 \oplus 45 \\
10 \otimes 10 &= \bar{5} \oplus \overline{45} \oplus \overline{50} \,.
\end{aligned} \tag{A.10}$$

Explicit tensor decompositions SO(10) tensor representations $\phi_{\mu\nu...}$ of arbitrary dimension can be decomposed into SU(5) fields following Ref. [43],[1]

$$\begin{aligned}
\phi_{c_a \, c_b ...} &= \phi_{2a-1 \, c_b ...} - i \phi_{2a \, c_b ...} \,, \\
\phi_{\bar{c}_a \, c_b ...} &= \phi_{2a-1 \, c_b ...} + i \phi_{2a \, c_b ...} \,,
\end{aligned} \tag{A.11}$$

---
[1] The formalism has been adapted to our conventions, cf. Eq. (1.64).

The decomposition of the tensor representation $45 = \phi_{\mu\nu}$ into reducible SU(5) representations is given by

$$\phi_{\mu\nu} = \frac{i^{\mu+\nu}}{4}\left(-\phi_{\bar{c}_a\bar{c}_b} + (-1)^\nu \phi_{\bar{c}_a c_b} + (-1)^\mu \phi_{c_a \bar{c}_b} - (-1)^{\mu+\nu}\phi_{c_a c_b}\right). \quad (A.12)$$

One identifies the irreducible SU(5) constituents of $45 = \phi_{\mu\nu} = 1_{45} \oplus 10_{45} \oplus \overline{10}_{45} \oplus 24_{45}$ [124],

$$\begin{aligned}\phi_{c_5\bar{c}_5} &= \sqrt{10}\,H\,, & \phi_{\bar{c}_a\bar{c}_b} &= \sqrt{2}\,H_{ab}\,, \\ \phi_{c_a c_b} &= \sqrt{2}\,H^{ab}\,, & \phi_{c_a\bar{c}_b} &= \sqrt{2}\,\Sigma^a_b + \tfrac{1}{5}\delta^a_b\sqrt{10}\,H\,,\end{aligned} \quad (A.13)$$

with

$$H = 1_{45}\,, \qquad H^{ab} = 10_{45}\,, \qquad H_{ab} = \overline{10}_{45}\,, \qquad \Sigma^a_b = 24_{45}\,. \quad (A.14)$$

A.3 Loop functions for meson mixing

The loop functions for meson mixing from the weak W box diagrams in the SM are given by the Inami-Lim functions [125]

$$S_0(x_c) = x_c, \quad (A.15)$$

$$S_0(x_t) = \frac{4x_t - 11x_t^2 + x_t^3}{4(1-x_t)^2} - \frac{3x_t^3 \log(x_t)}{2(1-x_t)^3}, \quad (A.16)$$

$$S_0(x_c, x_t) = x_c\left[\log\frac{x_t}{x_c} - \frac{3x_t}{4(1-x_t)} - \frac{3x_t^2 \log x_t}{4(1-x_t)^2}\right], \quad (A.17)$$

with $x_q = m_q^2/M_W^2$, where $m_q \equiv \overline{m}_q(m_q)$ is the quark mass in the renormalization scheme $\overline{\text{MS}}$. For the gluino box diagrams relevant in the CMM model, the loop functions read

$$L_0(x,y) = \frac{11}{18}G(x,y) - \frac{2}{9}F(x,y), \quad (A.18)$$

$$S^{(\tilde{g})}(x,y) = L_0(x,x) - 2L_0(x,y) + L_0(y,y), \quad (A.19)$$

with $F(x,y)$ and $G(x,y)$ defined in Ref. [126],

$$F(x,y) = -\frac{1}{(x-1)(y-1)} - \frac{1}{x-y}\left[\frac{x\ln x}{(x-1)^2} - \frac{y\ln y}{(y-1)^2}\right], \quad (A.20)$$

$$G(x,y) = \frac{1}{(x-1)(y-1)} + \frac{1}{x-y}\left[\frac{x^2\ln x}{(x-1)^2} - \frac{y^2\ln y}{(y-1)^2}\right]. \quad (A.21)$$

A.4 $B \to D$ form factors and decay distribution

Form factors Following the notations of Ref. [120], the hadronic form factors for $B \to D$ transitions can be parametrized by a power series in the kinematic variable z, cf. Eq. (4.27),[2]

$$F_j(t) = \frac{1}{P_j(t)\,\phi_j(t,t_0)} \sum_{k=0}^{\infty} a_k^j(t_0)\, z^k(t,t_0)\,, \qquad (A.22)$$

where $t_0 = t_+(1 - \sqrt{1 - t_-/t_+})$ has been chosen in order to minimize $|z^{\max}|$. For $B \to D$, one has $|z^{\max}| = 0.032$, such that it is sufficient to break the series after the linear term in z if the coefficients a_k are under control. This is ensured by introducing the functions $P(t)$ and $\phi(t)$. Resonances below the BD threshold $m_B + m_D \approx 7.15\,\text{GeV}$ are removed by

$$P(t) = \prod_i z(t, m_{B_{c,i}^*}^2 / m_B^2)\,. \qquad (A.23)$$

For the vector and scalar form factors, the relevant B_c^* resonances with $J^P = 1^-$ and 0^+ are [127]

$$\begin{aligned} P_V(t):&\quad m_{B_c^*} = 6.337,\ 6.899,\ 7.012\,\text{GeV}\,, \\ P_S(t):&\quad m_{B_c^*} = 6.700,\ 7.108\,\text{GeV}\,. \end{aligned} \qquad (A.24)$$

The functions $\phi_j(t)$ are taken by default as

$$\begin{aligned} \phi_V(t,t_0) &= \frac{1}{m_B}\frac{1}{\sqrt{24\pi}}\frac{t_+ - t}{(t_+ - t_0)^{1/4}} \left(\frac{z(t,0)}{-t}\right)^{5/2} \left(\frac{z(t,t_0)}{t_0 - t}\right)^{-1/2} \left(\frac{z(t,t_-)}{t_- - t}\right)^{-3/4}, \\ \phi_S(t,t_0) &= \sqrt{\frac{t_+ t_-}{8\pi}}\frac{\sqrt{t_+ - t}}{(t_+ - t_0)^{1/4}} \left(\frac{z(t,0)}{-t}\right)^{2} \left(\frac{z(t,t_0)}{t_0 - t}\right)^{-1/2} \left(\frac{z(t,t_-)}{t_- - t}\right)^{-1/4}, \end{aligned} \qquad (A.25)$$

setting $\eta = 2$ and $Q^2 = 0$ in Eq. (10) of Ref. [120]. The explicit expressions for F_V and F_S in terms of two parameters a_0 and a_1 are finally given by

$$\begin{aligned} F_V(t) &= \frac{1}{P_V(t)\,\phi_V(t,t_0)}\big\{a_0^V + a_1^V\, z(t,t_0)\big\}\,, \\ F_S(t) &= \frac{1}{P_S(t)\,\phi_S(t,t_0)}\big\{a_0^S + a_1^S\, z(t,t_0)\big\}\,. \end{aligned} \qquad (A.26)$$

The coefficients $a_{0,1}^V$ for the vector form factor are derived from a fit to the experimental spectrum of $B \to D\ell\nu$, cf. Eq. (4.29).[3] For the scalar form factor, $a_{0,1}^S$ are fixed by $F_S(0) = F_V(0)$ and $F_S(t_-) = \frac{2\sqrt{r}}{1+r}(1.02 \pm 0.05)$ from HQET. Using the inputs $|V_{cb}|$ from Tab. 4.1 and $|V_{cb}|\mathcal{G}(1)$, ρ^2 from Eq. (4.30), we derive the relevant coefficients for $B \to D$ form factor parametrizations given in Tab. A.1.

---

[2] Note that we rescaled the variable $t \to m_B^2 t$ with respect to Ref. [120], and accordingly for $t_{0,\pm}$.

[3] The parametrizations in Eqs. (4.29) and (A.26) are equivalent and result in the same shape for the vector form factor, when fitted to the spectrum.

A.4. B → D form factors and decay distribution

Parameters	centr. $|V_{cb}|F$	min. $|V_{cb}|F$	max. $|V_{cb}|F$
$\{|V_{cb}|\mathcal{G}(1), \rho^2\}$	$\{0.042, 1.18\}$	$\{0.041, 1.14\}$	$\{0.044, 1.22\}$
$|V_{cb}|\{a_0^V, a_1^V\}\,[10^{-5}]$	$\{1.09, -3.1\}$	$\{1.07, -2.6\}$	$\{1.12, -3.7\}$
$|V_{cb}|\{a_0^S, a_1^S\}\,[10^{-4}]$	$\{1.88, -6.9\}$	$\{1.80, -5.3\}$	$\{1.97, -8.5\}$

Table A.1: Parameter sets for the form factors $|V_{cb}|F_V$ and $|V_{cb}|F_S$. The results are shown in Fig. 4.5, where the envelopes of the error bands correspond to the inputs labeled by "min." and "max.".

Decay distribution The triple differential distribution in the decay chain $\bar{B} \to D\bar{\nu}_\tau \tau^-[\to \pi^- \nu_\tau]$ reads

$$\frac{d\Gamma(\bar{B} \to D\bar{\nu}_\tau \tau^-[\to \pi^- \nu_\tau])}{dE_D\, dE_\pi\, d\cos\theta_{D\pi}} = G_F^4 f_\pi^2 |V_{ud}|^2 |V_{cb}|^2 \tau_\tau \quad (A.27)$$

$$\times \left[C_W(F_V, F_S) - C_{WH}(F_V, F_S)\, \text{Re}[g_S] + C_H(F_S)|g_S|^2 \right].$$

Setting $m_\pi = 0$, which is good to 1% precision in $d^3\Gamma$, one has

$$C_W = \kappa \frac{m_\tau^4}{2} \frac{l^2}{p_\pi \cdot l} \left\{ P^2(b-1) + (P \cdot l)^2 \frac{2b}{l^2} + \left[\frac{l^2(P \cdot p_\pi)^2}{(p_\pi \cdot l)^2} - \frac{2(P \cdot l)(P \cdot p_\pi)}{p_\pi \cdot l} \right](3b-1) \right\},$$

$$C_{WH} = 2\kappa\, m_\tau^4 \frac{(1-r^2)F_S}{1 - \overline{m}_c/\overline{m}_b}\, b \left[P \cdot l - \frac{l^2 P \cdot p_\pi}{p_\pi \cdot l} \right],$$

$$C_H = \kappa\, m_\tau^6 \frac{(1-r^2)^2 F_S^2}{(1 - \overline{m}_c/\overline{m}_b)^2} \left(1 - \frac{m_\tau^2}{2 p_\pi \cdot l} \right),$$

(A.28)

with the abbreviations

$$P = F_V(p_B + p_D) - (F_V - F_S)\frac{m_B^2(1-r^2)}{q^2}(p_B - p_D),$$

$$\kappa = \frac{E_\pi \sqrt{E_D^2 - m_D^2}}{128\, \pi^4 m_B m_\tau}, \qquad b = \frac{m_\tau^2}{p_\pi \cdot l}\left(1 - \frac{m_\tau^2}{2\, p_\pi \cdot l}\right), \quad (A.29)$$

$$l = p_B - p_D - p_\pi, \qquad q^2 = (p_B - p_D)^2.$$

The dot products appearing in Eqs. (A.28) and (A.29) are related to the energies, momenta, and the angle $\theta_{D\pi}$ measured in the B rest frame as

$$p_B \cdot l = m_B(m_B - E_D - E_\pi), \qquad p_D \cdot l = E_D(m_B - E_D - E_\pi) + |\vec{p}_D|^2 + |\vec{p}_D|E_\pi \cos\theta_{D\pi},$$

$$p_\pi \cdot l = E_\pi(m_B - E_D) + |\vec{p}_D|E_\pi \cos\theta_{D\pi}, \qquad p_B \cdot p_D = m_B E_D. \quad (A.30)$$

Bibliography

[1] S. Trine, S. Westhoff, and S. Wiesenfeldt, "Probing Yukawa Unification with K and B Mixing," *JHEP* **08** (2009) 002, `arXiv:0904.0378 [hep-ph]`.

[2] U. Nierste, S. Trine, and S. Westhoff, "Charged-Higgs effects in a new $B \to D\tau\nu$ differential decay distribution," *Phys. Rev.* **D78** (2008) 015006, `arXiv:0801.4938 [hep-ph]`.

[3] S. Weinberg, "A Model of Leptons," *Phys. Rev. Lett.* **19** (1967) 1264.

[4] A. Salam, "Elementary Particle Theory," *ed. N. Svartholm, Almquist and Wiksells, Stockholm* (1969) 367.

[5] S. L. Glashow, J. Iliopoulos, and L. Maiani, "Weak Interactions with Lepton-Hadron Symmetry," *Phys. Rev.* **D2** (1970) 1285.

[6] P. W. Higgs, "Broken symmetries, massless particles and gauge fields," *Phys. Lett.* **12** (1964) 132.

[7] P. W. Higgs, "Broken symmetries and the masses of gauge bosons," *Phys. Rev. Lett.* **13** (1964) 508.

[8] F. Englert and R. Brout, "Broken symmetry and the mass of gauge vector mesons," *Phys. Rev. Lett.* **13** (1964) 321.

[9] G. S. Guralnik, C. R. Hagen, and T. W. B. Kibble, "Global conservation laws and massless particles," *Phys. Rev. Lett.* **13** (1964) 585.

[10] N. Cabibbo, "Unitary Symmetry and Leptonic Decays," *Phys. Rev. Lett.* **10** (1963) 531.

[11] M. Kobayashi and T. Maskawa, "CP Violation in the Renormalizable Theory of Weak Interaction," *Prog. Theor. Phys.* **49** (1973) 652.

[12] Z. Maki, M. Nakagawa, and S. Sakata, "Remarks on the unified model of elementary particles," *Prog. Theor. Phys.* **28** (1962) 870.

[13] B. Pontecorvo, "Neutrino experiments and the question of leptonic-charge conservation," *Sov. Phys. JETP* **26** (1968) 984.

[14] P. F. Harrison, D. H. Perkins, and W. G. Scott, "Tri-bi-maximal mixing and the neutrino oscillation data," *Phys. Lett.* **B530** (2002) 167, arXiv:hep-ph/0202074.

[15] H. Fritzsch and Z.-Z. Xing, "Lepton Mass Hierarchy and Neutrino Oscillations," *Phys. Lett.* **B372** (1996) 265, arXiv:hep-ph/9509389.

[16] P. Minkowski, "$\mu \to e\gamma$ at a Rate of One Out of 1-Billion Muon Decays?," *Phys. Lett.* **B67** (1977) 421.

[17] T. Yanagida, "Horizontal gauge symmetry and masses of neutrinos," *in Proceedings of the Workshop on the Baryon Number of the Universe and Unified Theories, Tsukuba, Japan, 13-14 Feb 1979*.

[18] M. Gell-Mann, P. Ramond, and R. Slansky, "Complex Spinors and Unified Theories," *in Supergravity, P. van Nieuwenhuizen & D.Z. Freedman (eds.), North Holland Publ. Co.* (1979).

[19] J. F. Gunion, H. E. Haber, G. Kane, and S. Dawson, "The Higgs Hunter's Guide," *Westview Press, Boulder, Colorado* (2000).

[20] J. R. Ellis and D. V. Nanopoulos, "Flavor Changing Neutral Interactions in Broken Supersymmetric Theories," *Phys. Lett.* **B110** (1982) 44.

[21] R. Barbieri and R. Gatto, "Conservation Laws for Neutral Currents in Spontaneously Broken Supersymmetric Theories," *Phys. Lett.* **B110** (1982) 211.

[22] A. J. Buras, P. Gambino, M. Gorbahn, S. Jäger, and L. Silvestrini, "ϵ'/ϵ and Rare K and B Decays in the MSSM," *Nucl. Phys.* **B592** (2001) 55, arXiv:hep-ph/0007313.

[23] G. D'Ambrosio, G. F. Giudice, G. Isidori, and A. Strumia, "Minimal flavour violation: An effective field theory approach," *Nucl. Phys.* **B645** (2002) 155, arXiv:hep-ph/0207036.

[24] C. Giunti, C. W. Kim, and U. W. Lee, "Running coupling constants and grand unification models," *Mod. Phys. Lett.* **A6** (1991) 1745.

[25] P. Langacker and M.-x. Luo, "Implications of precision electroweak experiments for M_t, ρ_0, $\sin^2\theta_W$ and grand unification," *Phys. Rev.* **D44** (1991) 817.

[26] U. Amaldi, W. de Boer, and H. Fürstenau, "Comparison of grand unified theories with electroweak and strong coupling constants measured at LEP," *Phys. Lett.* **B260** (1991) 447.

[27] H. Georgi and S. L. Glashow, "Unity of All Elementary Particle Forces," *Phys. Rev. Lett.* **32** (1974) 438.

[28] L. J. Hall, V. A. Kostelecky, and S. Raby, "New Flavor Violations in Supergravity Models," *Nucl. Phys.* **B267** (1986) 415.

[29] F. Borzumati and A. Masiero, "Large Muon and electron Number Violations in Supergravity Theories," *Phys. Rev. Lett.* **57** (1986) 961.

[30] T. Moroi, "CP violation in $B_d \to \phi K_S$ in SUSY GUT with right-handed neutrinos," *Phys. Lett.* **B493** (2000) 366, arXiv:hep-ph/0007328.

[31] T. Moroi, "Effects of the right-handed neutrinos on $\Delta S = 2$ and $\Delta B = 2$ processes in supersymmetric SU(5) model," *JHEP* **03** (2000) 019, arXiv:hep-ph/0002208.

[32] J. Hisano and Y. Shimizu, "CP Violation in B_s Mixing in the SUSY SU(5) GUT with Right-handed Neutrinos," *Phys. Lett.* **B669** (2008) 301, arXiv:0805.3327 [hep-ph].

[33] K. Cheung, S. K. Kang, C. S. Kim, and J. Lee, "Correlation between lepton flavor violation and $B_{d,s} - \overline{B}_{d,s}$ mixing in SUSY GUT," *Phys. Lett.* **B652** (2007) 319, arXiv:hep-ph/0702050.

[34] H. Fritzsch and P. Minkowski, "Unified Interactions of Leptons and Hadrons," *Ann. Phys.* **93** (1975) 193.

[35] H. Georgi, "The State of the Art - Gauge Theories. (Talk)," *AIP Conf. Proc.* **23** (1975) 575.

[36] D. Chang, A. Masiero, and H. Murayama, "Neutrino mixing and large CP violation in B physics," *Phys. Rev.* **D67** (2003) 075013, arXiv:hep-ph/0205111.

[37] S. Jäger and U. Nierste, "B_s - $\overline{B}_s$ mixing in an SO(10) SUSY GUT model," *Eur. Phys. J.* **C33** (2004) 256, arXiv:hep-ph/0312145.

[38] R. Harnik, D. T. Larson, H. Murayama, and A. Pierce, "Atmospheric neutrinos can make beauty strange," *Phys. Rev.* **D69** (2004) 094024, arXiv:hep-ph/0212180.

[39] H. Georgi, "Lie Algebras in Particle Physics," *Perseus Books* (1999) .

[40] F. Wilczek and A. Zee, "Families from Spinors," *Phys. Rev.* **D25** (1982) 553.

[41] J. C. Pati and A. Salam, "Lepton Number as the Fourth Color," *Phys. Rev.* **D10** (1974) 275.

[42] C. S. Aulakh and R. N. Mohapatra, "Implications of supersymmetric SO(10) Grand Unification," *Phys. Rev.* **D28** (1983) 217.

[43] P. Nath and R. M. Syed, "Analysis of couplings with large tensor representations in SO(2N) and proton decay," *Phys. Lett.* **B506** (2001) 68, arXiv:hep-ph/0103165.

[44] S. Bertolini, M. Frigerio, and M. Malinsky, "Fermion masses in SUSY SO(10) with type II seesaw: A non-minimal predictive scenario," *Phys. Rev.* **D70** (2004) 095002, arXiv:hep-ph/0406117.

[45] D. Chang, T. Fukuyama, Y.-Y. Keum, T. Kikuchi, and N. Okada, "Perturbative SO(10) grand unification," *Phys. Rev.* **D71** (2005) 095002, arXiv:hep-ph/0412011.

[46] S. Wiesenfeldt, "Operator analysis for proton decay in SUSY SO(10) GUT models," *Phys. Rev.* **D71** (2005) 075006, arXiv:hep-ph/0501223.

[47] S. Jäger, M. Knopf, W. Martens, U. Nierste, C. Scherrer, and S. Wiesenfeldt. In preparation.

[48] S. P. Martin and M. T. Vaughn, "Two loop renormalization group equations for soft supersymmetry breaking couplings," *Phys. Rev.* **D50** (1994) 2282, arXiv:hep-ph/9311340.

[49] J. Hisano and D. Nomura, "Solar and atmospheric neutrino oscillations and lepton flavor violation in supersymmetric models with the right-handed neutrinos," *Phys. Rev.* **D59** (1999) 116005, arXiv:hep-ph/9810479.

[50] M. Drees, "Intermediate Scale Symmetry Breaking and the Spectrum of Super Partners in Superstring Inspired Supergravity Models," *Phys. Lett.* **B181** (1986) 279.

[51] M. Drees and S. P. Martin, "Implications of SUSY model building," arXiv:hep-ph/9504324.

[52] W. Martens. Private communication.

[53] J. A. Casas, "Charge and color breaking," arXiv:hep-ph/9707475. See also references therein.

[54] **Particle Data Group** Collaboration, C. Amsler et al., "Review of particle physics," *Phys. Lett.* **B667** (2008) 1.

[55] **LEP Working Group for Higgs boson searches** Collaboration, R. Barate et al., "Search for the standard model Higgs boson at LEP," *Phys. Lett.* **B565** (2003) 61, arXiv:hep-ex/0306033.

[56] J. R. Ellis, G. Ridolfi, and F. Zwirner, "Radiative corrections to the masses of supersymmetric Higgs bosons," *Phys. Lett.* **B257** (1991) 83.

[57] Y. Okada, M. Yamaguchi, and T. Yanagida, "Upper bound of the lightest Higgs boson mass in the minimal supersymmetric standard model," *Prog. Theor. Phys.* **85** (1991) 1.

[58] H. E. Haber and R. Hempfling, "Can the mass of the lightest Higgs boson of the minimal supersymmetric model be larger than M_Z?," *Phys. Rev. Lett.* **66** (1991) 1815.

[59] V. Weisskopf and E. P. Wigner, "Calculation of the natural brightness of spectral lines on the basis of Dirac's theory," *Z. Phys.* **63** (1930) 54.

BIBLIOGRAPHY

[60] V. Weisskopf and E. Wigner, "Over the natural line width in the radiation of the harmonius oscillator," *Z. Phys.* **65** (1930) 18.

[61] K. Anikeev *et al.*, "B physics at the Tevatron: Run II and beyond," arXiv:hep-ph/0201071.

[62] I. Bigi and A. Sanda, "CP Violation," *Cambridge University Press, UK* (2000).

[63] **Heavy Flavor Averaging Group** Collaboration, E. Barberio *et al.*, "Averages of b–hadron and c–hadron Properties at the End of 2007," arXiv:0808.1297 [hep-ex]. Updates on http://www.slac.stanford.edu/xorg/hfag/.

[64] A. J. Buras, "Weak Hamiltonian, CP violation and rare decays," arXiv:hep-ph/9806471.

[65] S. Herrlich and U. Nierste, "The Complete $\Delta S = 2$ Hamiltonian in the Next-To-Leading Order," *Nucl. Phys.* **B476** (1996) 27, arXiv:hep-ph/9604330.

[66] A. J. Buras, M. Jamin, and P. H. Weisz, "Leading and next-to-leading QCD corrections to epsilon parameter and B^0 - $\bar{B}^0$ mixing in the presence of a heavy top quark," *Nucl. Phys.* **B347** (1990) 491.

[67] A. J. Buras and D. Guadagnoli, "Correlations among new CP violating effects in $\Delta F = 2$ observables," *Phys. Rev.* **D78** (2008) 033005, arXiv:0805.3887 [hep-ph].

[68] **FlaviaNet Working Group on Kaon Decays** Collaboration, M. Antonelli *et al.*, "Precision tests of the Standard Model with leptonic and semileptonic kaon decays," arXiv:0801.1817 [hep-ph]. http://www.lnf.infn.it/wg/vus/.

[69] V. Lubicz and C. Tarantino, "Flavour physics and Lattice QCD: averages of lattice inputs for the Unitarity Triangle Analysis," *Nuovo Cim.* **123B** (2008) 674, arXiv:0807.4605 [hep-lat].

[70] M. Battaglia *et al.*, "The CKM matrix and the unitarity triangle," *Workshop, CERN, Geneva, Switzerland, 13-16 Feb 2002: Proceedings* (2003), arXiv:hep-ph/0304132.

[71] **HPQCD** Collaboration, I. Allison *et al.*, "High-Precision Charm-Quark Mass from Current-Current Correlators in Lattice and Continuum QCD," *Phys. Rev.* **D78** (2008) 054513, arXiv:0805.2999 [hep-lat].

[72] **Tevatron Electroweak Working Group** Collaboration, "Combination of CDF and D0 Results on the Mass of the Top Quark," arXiv:0808.1089 [hep-ex].

[73] K. G. Chetyrkin, J. H. Kühn, and M. Steinhauser, "RunDec: A Mathematica package for running and decoupling of the strong coupling and quark masses," *Comput. Phys. Commun.* **133** (2000) 43, arXiv:hep-ph/0004189.

[74] G. Buchalla, A. J. Buras, and M. E. Lautenbacher, "Weak decays beyond leading logarithms," *Rev. Mod. Phys.* **68** (1996) 1125, arXiv:hep-ph/9512380.

[75] J. H. Kühn and M. Steinhauser, "A theory driven analysis of the effective QED coupling at M_Z," *Phys. Lett.* **B437** (1998) 425, arXiv:hep-ph/9802241.

[76] **CKMfitter Group** Collaboration, J. Charles *et al.*, "CP violation and the CKM matrix: Assessing the impact of the asymmetric B factories," *Eur. Phys. J.* **C41** (2005) 1, arXiv:hep-ph/0406184. Updates on http://ckmfitter.in2p3.fr.

[77] A. Lenz and U. Nierste, "Theoretical update of $B_s - \bar{B}_s$ mixing," *JHEP* **06** (2007) 072, arXiv:hep-ph/0612167.

[78] J. R. Ellis and M. K. Gaillard, "Fermion Masses and Higgs Representations in SU(5)," *Phys. Lett.* **B88** (1979) 315.

[79] B. Bajc, P. Fileviez Perez, and G. Senjanovic, "Proton decay in minimal supersymmetric SU(5)," *Phys. Rev.* **D66** (2002) 075005, arXiv:hep-ph/0204311.

[80] D. Emmanuel-Costa and S. Wiesenfeldt, "Proton decay in a consistent supersymmetric SU(5) GUT model," *Nucl. Phys.* **B661** (2003) 62, arXiv:hep-ph/0302272.

[81] E. Lunghi and A. Soni, "Possible Indications of New Physics in B_d-mixing and in $\sin(2\beta)$ Determinations," *Phys. Lett.* **B666** (2008) 162, arXiv:0803.4340 [hep-ph].

[82] A. J. Buras and D. Guadagnoli, "On the consistency between the observed amount of CP violation in the K^- and B_d systems within minimal flavor violation," arXiv:0901.2056 [hep-ph].

[83] E. Lunghi and A. Soni, "Hints for the scale of new CP-violating physics from B-CP anomalies," arXiv:0903.5059 [hep-ph].

[84] L. J. Hall, R. Rattazzi, and U. Sarid, "The Top quark mass in supersymmetric SO(10) unification," *Phys. Rev.* **D50** (1994) 7048, arXiv:hep-ph/9306309.

[85] T. Blazek, S. Raby, and S. Pokorski, "Finite supersymmetric threshold corrections to CKM matrix elements in the large $\tan\beta$ regime," *Phys. Rev.* **D52** (1995) 4151, arXiv:hep-ph/9504364.

[86] R. D. Peccei and H. R. Quinn, "CP Conservation in the Presence of Instantons," *Phys. Rev. Lett.* **38** (1977) 1440.

[87] R. D. Peccei and H. R. Quinn, "Constraints Imposed by CP Conservation in the Presence of Instantons," *Phys. Rev.* **D16** (1977) 1791.

[88] L. Hofer, U. Nierste, and D. Scherer, "Resummation of $\tan\beta$-enhanced supersymmetric loop corrections beyond the decoupling limit," arXiv:0907.5408 [hep-ph].

[89] M. S. Carena, D. Garcia, U. Nierste, and C. E. M. Wagner, "Effective Lagrangian for the $\bar{t}bH^+$ interaction in the MSSM and charged Higgs phenomenology," *Nucl. Phys.* **B577** (2000) 88, arXiv:hep-ph/9912516.

[90] C. Hamzaoui, M. Pospelov, and M. Toharia, "Higgs-mediated FCNC in supersymmetric models with large $\tan\beta$," *Phys. Rev.* **D59** (1999) 095005, arXiv:hep-ph/9807350.

[91] A. Dedes and A. Pilaftsis, "Resummed effective Lagrangian for Higgs-mediated FCNC interactions in the CP-violating MSSM," *Phys. Rev.* **D67** (2003) 015012, arXiv:hep-ph/0209306.

[92] K. S. Babu and C. F. Kolda, "Higgs mediated $B^0 \to \mu^+\mu^-$ in minimal supersymmetry," *Phys. Rev. Lett.* **84** (2000) 228, arXiv:hep-ph/9909476.

[93] A. J. Buras, P. H. Chankowski, J. Rosiek, and L. Slawianowska, "$\Delta M_{d,s}$, $B^0_{d,s} \to \mu^+\mu^-$, and $B \to X_s\gamma$ in supersymmetry at large $\tan\beta$," *Nucl. Phys.* **B659** (2003) 3, arXiv:hep-ph/0210145.

[94] M. S. Carena, D. Garcia, U. Nierste, and C. E. M. Wagner, "$b \to s\gamma$ and supersymmetry with large $\tan\beta$," *Phys. Lett.* **B499** (2001) 141, arXiv:hep-ph/0010003.

[95] D0 Collaboration and V. Abazov, "Search for charged Higgs bosons in top quark decays," arXiv:0908.1811 [hep-ex].

[96] B. Dudley and C. Kolda, "Constraining the Charged Higgs Mass in the MSSM: A Low-Energy Approach," arXiv:0901.3337 [hep-ph].

[97] F. Domingo and U. Ellwanger, "Updated Constraints from B Physics on the MSSM and the NMSSM," *JHEP* **12** (2007) 090, arXiv:0710.3714 [hep-ph].

[98] G. Isidori and P. Paradisi, "Hints of large $\tan\beta$ in flavour physics," *Phys. Lett.* **B639** (2006) 499, arXiv:hep-ph/0605012.

[99] **BABAR** Collaboration, B. Aubert *et al.*, "Measurement of $|V_{cb}|$ and the Form-Factor Slope in $\overline{B} \to D\ell^-\bar{\nu}$ Decays in Events Tagged by a Fully Reconstructed B Meson," arXiv:0904.4063 [hep-ex].

[100] **Belle** Collaboration, K. Abe *et al.*, "Measurement of $\mathcal{B}(\overline{B} \to D^+\ell^-\bar{\nu})$ and determination of $|V_{cb}|$," *Phys. Lett.* **B526** (2002) 258, arXiv:hep-ex/0111082.

[101] B. Grzadkowski and W.-S. Hou, "Searching for $B \to D\tau\bar{\nu}$ at the 10-percent level," *Phys. Lett.* **B283** (1992) 427.

[102] S. Marchetti, S. Mertens, U. Nierste, and D. Stöckinger, "$\tan\beta$-enhanced supersymmetric corrections to the anomalous magnetic moment of the muon," *Phys. Rev.* **D79** (2009) 013010, arXiv:0808.1530 [hep-ph].

[103] W.-S. Hou, "Enhanced charged Higgs boson effects in $B^- \to \tau\bar{\nu}$, $\mu\bar{\nu}$, and $b \to \tau\bar{\nu} + X$," *Phys. Rev.* **D48** (1993) 2342.

[104] **CKMfitter** Collaboration. Fit inputs for winter 2009. Updates on http://www.slac.stanford.edu/xorg/ckmfitter/.

[105] K. G. Chetyrkin *et al.*, "Charm and Bottom Quark Masses: an Update," arXiv:0907.2110 [hep-ph].

[106] S. Bethke, "Experimental tests of asymptotic freedom," *Prog. Part. Nucl. Phys.* **58** (2007) 351, arXiv:hep-ex/0606035.

[107] K. Kiers and A. Soni, "Improving constraints on $\tan\beta/m_H$ using $B \to D\tau\bar{\nu}$," *Phys. Rev.* **D56** (1997) 5786, arXiv:hep-ph/9706337.

[108] A. V. Manohar and M. B. Wise, "Heavy Quark Physics," *Cambridge University Press, UK* (2000).

[109] **BABAR** Collaboration, B. Aubert *et al.*, "Measurement of the Semileptonic Decays $B \to D\tau^-\bar{\nu}_\tau$ and $B \to D^*\tau^-\bar{\nu}_\tau$," *Phys. Rev.* **D79** (2009) 092002, arXiv:0902.2660 [hep-ex].

[110] T. Miki, T. Miura, and M. Tanaka, "Effects of charged Higgs boson and QCD corrections in $\bar{B} \to D\tau\bar{\nu}$," arXiv:hep-ph/0210051.

[111] B. K. Bullock, K. Hagiwara, and A. D. Martin, "Tau polarization as a signal of charged Higgs bosons," *Phys. Rev. Lett.* **67** (1991) 3055–3057.

[112] N. Isgur and M. B. Wise, "Weak Decays of Heavy Mesons in the Static Quark Approximation," *Phys. Lett.* **B232** (1989) 113. And references therein.

[113] B. Grinstein, "The static quark effective theory," *Nucl. Phys.* **B339** (1990) 253.

[114] E. Eichten and B. R. Hill, "Static effective field theory: $1/m$ corrections," *Phys. Lett.* **B243** (1990) 427.

[115] H. Georgi, "An effective field theory for heavy quarks at low energies," *Phys. Lett.* **B240** (1990) 447.

[116] N. Isgur and M. B. Wise, "Weak transition form factors between heavy mesons," *Phys. Lett.* **B237** (1990) 527.

[117] M. E. Luke, "Effects of subleading operators in the heavy quark effective theory," *Phys. Lett.* **B252** (1990) 447.

[118] M. Neubert, "Short distance expansion of heavy quark currents," *Phys. Rev.* **D46** (1992) 2212.

[119] I. Caprini, L. Lellouch, and M. Neubert, "Dispersive bounds on the shape of $\overline{B} \to D^{(*)}\ell\bar{\nu}$ form factors," *Nucl. Phys.* **B530** (1998) 153, arXiv:hep-ph/9712417.

[120] R. J. Hill, "The modern description of semileptonic meson form factors," arXiv:hep-ph/0606023.

[121] G. M. de Divitiis, R. Petronzio, and N. Tantalo, "Quenched lattice calculation of semileptonic heavy-light meson form factors," *JHEP* **10** (2007) 062, arXiv:0707.0587 [hep-lat].

[122] J. F. Kamenik and F. Mescia, "$B \to D\tau\nu$ Branching Ratios: Opportunity for Lattice QCD and Hadron Colliders," *Phys. Rev.* **D78** (2008) 014003, arXiv:0802.3790 [hep-ph].

[123] B. Ananthanarayan, K. S. Babu, and Q. Shafi, "Supersymmetric models with $\tan\beta$ close to unity," *Nucl. Phys.* **B428** (1994) 19, arXiv:hep-ph/9402284.

[124] P. Nath and R. M. Syed, "Complete cubic and quartic couplings of 16 and $\overline{16}$ in SO(10) unification," *Nucl. Phys.* **B618** (2001) 138, arXiv:hep-th/0109116.

[125] T. Inami and C. S. Lim, "Effects of Superheavy Quarks and Leptons in Low-Energy Weak Processes $K_L \to \mu^+\mu^-$, $K^+ \to \pi^+\nu\bar{\nu}$, and $K^0 \leftrightarrow \overline{K}^0$," *Prog. Theor. Phys.* **65** (1981) 297. [Erratum: **65** (1981) 1772].

[126] S. Bertolini, F. Borzumati, A. Masiero, and G. Ridolfi, "Effects of supergravity induced electroweak breaking on rare B decays and mixings," *Nucl. Phys.* **B353** (1991) 591.

[127] E. J. Eichten and C. Quigg, "Mesons with beauty and charm: Spectroscopy," *Phys. Rev.* **D49** (1994) 5845, arXiv:hep-ph/9402210.

Acknowledgements

- I thank Prof. Ulrich Nierste for supervising my work, for interesting discussions and helpful explanations, and for supporting me as a novice in the physicists' community. His animated way of talking about particle physics has always been inspiring to me.
- I thank "my" postdoc Stéphanie Trine for our close collaboration. I am greatful for her constant encouragements and for having adapted part of her thoroughness in doing research.
- I thank Sören Wiesenfeldt for our collaboration, especially for his patience when introducing me to the fundamentals of Grand Unification.
- I thank all my colleagues and in particular my "roommates" for good discussions and for their readiness to help with computer issues. I had an enjoyable time in our institute at work and beyond.
- I thank Prof. Matthias Steinhauser for being the second examiner of my thesis.
- I thank Stefan Bekavac, Jennifer Girrbach, Waldemar Martens, and Sören Wiesenfeldt for proofreading the manuscript and for helpful comments.
- I thank my good, trusty family and all those who accompanied me, in particular Argelia Rodriguez and P. Victor Diaz Aleman. I am very greatful for their support.

This work has been financially supported by the DFG-SFB/TR9 and by the DFG "Graduiertenkolleg Hochenergiephysik und Astroteilchenphysik".

Die VDM Verlagsservicegesellschaft sucht für wissenschaftliche Verlage abgeschlossene und herausragende

Dissertationen, Habilitationen, Diplomarbeiten, Master Theses, Magisterarbeiten usw.

für die kostenlose Publikation als Fachbuch.

Sie verfügen über eine Arbeit, die hohen inhaltlichen und formalen Ansprüchen genügt, und haben Interesse an einer honorarvergüteten Publikation?

Dann senden Sie bitte erste Informationen über sich und Ihre Arbeit per Email an *info@vdm-vsg.de*.

Sie erhalten kurzfristig unser Feedback!

VDM Verlagsservicegesellschaft mbH
Dudweiler Landstr. 99   Telefon +49 681 3720 174
D - 66123 Saarbrücken   Fax   +49 681 3720 1749
www.vdm-vsg.de

Die VDM Verlagsservicegesellschaft mbH vertritt

Printed by Books on Demand GmbH, Norderstedt / Germany